Graduate Texts in Contemporary Physics

Series Editors:

Joseph L. Birman
Jeffrey W. Lynn
Mark P. Silverman
H. Eugene Stanley
Mikhail Voloshin

Springer
New York
Berlin
Heidelberg
Barcelona
Hong Kong
London
Milan
Paris
Singapore
Tokyo

Graduate Texts in Contemporary Physics

A. Auerbach: **Interacting Electrons and Quantum Magnetism**

B. Felsager: **Geometry, Particles, and Fields**

P. Di Francesco, P. Mathieu, and D. Sénéchal: **Conformal Field Theories**

J.H. Hinken: **Superconductor Electronics: Fundamentals and Microwave Applications**

J. Hladik: **Spinors in Physics**

Yu.M. Ivanchenko and A.A. Lisyansky: **Physics of Critical Fluctuations**

M. Kaku: **Introduction to Superstrings and M-Theory, 2nd Edition**

M. Kaku: **Strings, Conformal Fields, and Topology: An Introduction**

H.V. Klapdor (ed.): **Neutrinos**

J.W. Lynn (ed.): **High-Temperature Superconductivity**

H.J. Metcalf and P. van der Straten: **Laser Cooling and Trapping**

R.N. Mohapatra: **Unification and Supersymmetry: The Frontiers of Quark-Lepton Physics, 2nd Edition**

H. Oberhummer: **Nuclei in the Cosmos**

R.E. Prange and S.M. Girvin (eds.): **The Quantum Hall Effect**

F.T. Vasko and A.V. Kuznetsov: **Electronic States and Optical Transitions in Semiconductor Heterostructures**

A.M. Zagoskin: **Quantum Theory of Many-Body Systems: Techniques and Applications**

Jean Hladik

Spinors in Physics

Translated by J. Michael Cole

Jean Hladik
Department of Physics
University of Angers
49035 Angers
France

With 7 figures.

Library of Congress Cataloging-in-Publication Data
Hladik, Jean.
[Spineurs en physique. English]
Spinors in physics/Jean Hladik; translated by J. Michael Cole.
p. cm. — (Graduate texts in contemporary physics)
Includes bibliographical references and index.
ISBN 0-387-98647-2 (hardcover: alk. paper)
1. Spinor analysis. 2. Mathematical physics. I. Title.
II. Series.
QC20.7.S65H5513 1999
530.15′683—dc21 99-24764

Printed on acid-free paper.

"Les spineurs en physique" by Jean Hladik. © Masson, Paris 1996
Translated from French by J. Michael Cole.

Production managed by Frank McGuckin; manufacturing supervised by Jeffrey Taub.
Photocomposed copy prepared from the author's TeX files.
Printed and bound by R.R. Donnelley and Sons, Harrisonburg, VA.
Printed in the United States of America.

9 8 7 6 5 4 3 2 1

ISBN 0-387-98647-2 Springer-Verlag New York Berlin Heidelberg SPIN 10696772

Preface

Spinors play a major role in quantum mechanics and, consequently, in the whole of contemporary physics. It was at almost the beginning of quantum theory, in 1927, that the physicists Pauli, and then Dirac, introduced spinors to represent wave functions; the former for three-dimensional space and the latter for four-dimensional space–time.

However, in their mathematical form spinors had been discovered by Élie Cartan in 1913 during his research into the representation of groups. He showed then that spinors furnish a linear representation of the groups of rotations of a space of arbitrary dimension.

Spinors are thus directly related to geometry, but they are often presented in an abstract way that does not have any intuitive geometric meaning. Moreover, it was this criticism that Élie Cartan made in the introduction of his book *Leçons sur la théorie des spineurs* [3] where he wrote, "One of the principal goals of this work is to develop the theory of spinors systematically by giving these mathematical entities a purely geometric definition: thanks to their geometric origin, the matrices used in quantum mechanics by physicists are themselves presented."

This geometric approach to spinors is also that used by Max Morand in his book *Géométrie spinorielle* [16]. As we have also had occasion to say—Professor Max Morand directed the laboratory where I prepared my Doctor of Science thesis—it was in the course of conversations with Élie Cartan in the buildings of the École Normale Supérieure that Max Morand became convinced that a geometric presentation of spinors was the best approach for physicists. Without doubt it is this influence which has encouraged me to begin this book with the geometry of two-component spinors, but I am also persuaded that it is the best way for a physicist to connect certain quantum phenomena with their mathematical depiction.

After a first, simple, geometric approach to spinors we have used the formalism of Lie group theory to recover spinors and the irreducible representations of these groups.

If the theory of Lie groups is difficult and is not generally known by undergraduate physics students, it is, however, possible, as we have also done, to give some outlines of how to use them for our needs. By restricting ourselves to the linear groups and the most classic ones, and using their representations in matrix form we easily obtain the information that suffices for the needs of an undergraduate course in quantum mechanics where spinors are introduced. Now, for it to be properly assimilated the notion of spinor demands, we suggest, a minimum of an undergraduate course on the theory of Lie groups. We have already advocated this in a previous book [11] devoted to the application of the theory of groups to quantum mechanics.

The spinor formalism is not of interest solely in quantum mechanics, for the works, amongst others, of Roger Penrose have shown that spinor theory is an extremely productive approach in the relativistic theory of gravitation. Although the formalism currently used the most widely for General Relativity is tensor calculus, that particular author has shown that in the specific case of four-dimensional space and the Lorentzian metric the formalism of two-component spinors is the most appropriate.

Our most grateful thanks go to our colleague Claude Latrémolière, professor of mathematics at the University of Angers, for his careful reading of our manuscript as well as for his remarks and advice which have enabled us to make a much improved presentation of the contents of this book.

Contents

Preface v

I Spinors in Three-Dimensional Space 1

1 Two-Component Spinor Geometry 3
1.1 Definition of a Spinor 4
1.1.1 Stereographic Projection 4
1.1.2 Vectors Associated with a Spinor 5
1.1.3 The Definition of a Spinor 7
1.2 Geometrical Properties 9
1.2.1 Plane Symmetries 9
1.2.2 Rotations 11
1.2.3 The Olinde–Rodrigues Parameters 14
1.2.4 Rotations Defined in Terms of the Euler Angles 15
1.3 Infinitesimal Properties of Rotations 17
1.3.1 The Infinitesimal Rotation Matrix 17
1.3.2 The Pauli Matrices 18
1.3.3 Properties of the Pauli Matrices 20
1.4 Algebraic Properties of Spinors 22
1.4.1 Operations on Spinors 22
1.4.2 Properties of Operations on Spinors 23
1.4.3 The Basis of the Vector Space of Spinors 24
1.4.4 Hermitian Vector Spaces 25
1.4.5 Properties of the Hermitian Product 26
1.4.6 The Use of an Antisymmetric Metric Tensor 28
1.5 Solved Problems 29

2 Spinors and $SU(2)$ Group Representations 35
2.1 Lie Groups 35
2.1.1 Examples of Continuous Groups 35

2.1.2 Analytic Definition of Continuous Groups 37
2.1.3 Linear Representations 39
2.1.4 Infinitesimal Generators 41
2.1.5 Infinitesimal Matrices 43
2.1.6 Exponential Mapping 46
2.1.7 The Nomenclature of Continuous Linear Groups . . 47
2.2 Unimodular Unitary Groups 48
2.2.1 The Unitary Group $\boldsymbol{U}(2)$ 48
2.2.2 The Unitary Unimodular Group $\boldsymbol{SU}(2)$ 50
2.2.3 Three-Dimensional Representations 52
2.2.4 Representations of the Groups $\boldsymbol{SU}(2)$ 53
2.2.5 Irreducible Representations of $\boldsymbol{SU}(2)$ 55
2.3 Solved Problems 57

3 Spinor Representation of $\boldsymbol{SO}(3)$ **67**
3.1 The Rotation Group $\boldsymbol{SO}(3)$ 67
3.1.1 Rotations About a Point 67
3.1.2 The Infinitesimal Matrices of the Group 68
3.1.3 Rotations About a Given Axis 70
3.1.4 The Exponential Matrix of a Rotation About a Given Axis 72
3.2 Irreducible Representations of $\boldsymbol{SO}(3)$ 73
3.2.1 The Structure Equations 73
3.2.2 The Infinitesimal Matrices of the Representations of the Group $\boldsymbol{SO}(3)$ 75
3.2.3 Eigenvectors and Eigenvalues of the Infinitesimal Matrices of the Representations 76
3.2.4 Irreducible Representations 78
3.2.5 The Infinitesimal Matrices of an Irreducible Representation in the Canonical Basis 79
3.2.6 The Characters of the Rotation Matrices of a Representation 81
3.3 Spherical Harmonics 82
3.3.1 The Infinitesimal Operators in Spherical Coordinates 82
3.3.2 Spherical Harmonics 83
3.4 Spinor Representations 85
3.4.1 The Two-Dimensional Irreducible Representation . . 85
3.4.2 The Three-Dimensional Irreducible Representation . 87
3.4.3 $(2j+1)$-Dimensional Irreducible Representations . . 91
3.5 Solved Problems 93

4 Pauli Spinors **99**
4.1 Spin and Spinors 99
4.2 The Linearized Schrödinger Equations 100

4.2.1 The Free Particle 100
4.2.2 Particle in an Electromagnetic Field 104
4.2.3 The Spinors in Pauli's Equation 105
4.3 Spinor and Vector Fields 108
4.3.1 The Transformation of a Vector Field by a Rotation 108
4.3.2 The Rotation of a Spinor Field 110
4.4 Solved Problems 112

II Spinors in Four-Dimensional Space 119

5 The Lorentz Group 121
5.1 The Generalized Lorentz Group 121
5.1.1 Rotations and Reflections 121
5.1.2 Orthochronous and Anti-Orthochronous Transformations 123
5.1.3 Sheets of the Generalized Lorentz Group 123
5.2 The Four-Dimensional Rotation Group 125
5.2.1 Four-Dimensional Orthogonal Transformations . . . 125
5.2.2 Matrix Representations of the Group $\boldsymbol{SO}(4)$ 126
5.2.3 Infinitesimal Matrices 128
5.2.4 Irreducible Representations 130
5.3 Solved Problems 131

6 Representations of the Lorentz Groups 135
6.1 Irreducible Representations 135
6.1.1 Relations Between the Groups $\boldsymbol{SO}(3,1)^{\uparrow}$ and $\boldsymbol{SO}(4)$ 135
6.1.2 Infinitesimal Matrices 137
6.1.3 Irreducible Representations 139
6.2 The Group $\boldsymbol{SL}(2,\mathbb{C})$ 140
6.2.1 Two-Component Spinors 140
6.2.2 Higher-Order Spinors 141
6.2.3 Representations of the Groups $\boldsymbol{SL}(2,\mathbb{C})$ 142
6.2.4 Irreducible Representations 144
6.3 Spinor Representations of the Lorentz Group 145
6.3.1 Four-Dimensional Irreducible Representations 145
6.3.2 Two-Dimensional Representations 147
6.3.3 The Direct Product of Irreducible Representations . 149
6.4 Solved Problems 150

7 Dirac Spinors 157
7.1 The Dirac Equation 157
7.1.1 The Classical Relativistic Wave Equation 157
7.1.2 The Dirac Equation for a Free Particle 158
7.1.3 A Particle in an Electromagnetic Field 159

7.2 Relativistic Invariance of the Dirac Equation 160
7.2.1 The Relativistic Invariance Condition 160
7.2.2 The Type of Representation for the Wave Function . 161
7.2.3 The Link Between a Spinor and a Four-Vector . . . 162
7.2.4 Dirac's Equation in the Spinor Representation . . . 164
7.2.5 The Symmetric Form of the Dirac Equation 165
7.3 Solved Problems 166

8 Clifford and Lie Algebras 171
8.1 Lie Algebras 171
8.1.1 The Definition of an Algebra 171
8.1.2 Lie Algebras 172
8.1.3 Isomorphic Lie Algebras 173
8.2 Representations of Lie Algebras 174
8.2.1 Definition 174
8.2.2 Representations of a Lie Group and of Its Lie Algebra 176
8.2.3 Connected Groups 178
8.2.4 Reducible and Irreducible Representations 178
8.3 Clifford Algebras 179
8.3.1 Definition 179
8.3.2 Examples of Clifford Algebras 180
8.3.3 Clifford and Lie Algebras 184
8.3.4 Spinor Groups 186
8.4 Solved Problems 189

Appendix: Groups and Their Representations 197
A.1 The Definition of a Group 197
A.1.1 Examples of Groups 197
A.1.2 The Axioms Defining a Group 200
A.1.3 Elementary Properties of Groups 201
A.2 Linear Operators 203
A.2.1 The Operator Representing an Element of a Group . 203
A.2.2 The Operators Acting on the Vectors of Geometric Space 204
A.2.3 The Operators Acting on Wave Functions 205
A.2.4 Operators Representing a Group 207
A.3 Matrix Representations 207
A.3.1 The Rotation Matrix Acting on the Vectors of a Three-Dimensional Space 207
A.3.2 The Matrix of an Operator Acting on Functions . . 208
A.3.3 The Matrices Representing the Elements of a Group 209
A.4 Matrix Representations 210
A.4.1 The Definition of a Matrix Representation 210

A.4.2 The Fundamental Property of the Matrices of a Representation 210
A.4.3 Representation by Regular Matrices 211
A.4.4 Equivalent Representations 212
A.5 Reducible and Irreducible Representations 214
A.5.1 The Direct Sum of Two Vector Spaces 214
A.5.2 The Direct Sum of Two Representations 214
A.5.3 Irreducible Representations 215
A.6 The Direct Product of Representations 216
A.6.1 The Direct Product of Two Matrices 216
A.6.2 Properties of Tensor Products of Matrices 217
A.6.3 The Direct Product of Two Representations 219

References **221**

Index **223**

Part I

Spinors in Three-Dimensional Space

1

Two-Component Spinor Geometry

The term *spin* was introduced by physicists to describe certain properties of quantum particles which came to light in the course of various experiments. In order to describe these properties quantitatively some new mathematical entities were invented which were called *spinors*. These latter are vectors of a space the transformations of which are related in a particular way to rotations in physical space.

It was Pauli who, in 1927, put forward the idea that the wave function of an electron could be represented by a vector with two complex components, this latter being a spinor in three-dimensional Euclidean space. Then in the following year it was Dirac who created, for the needs of his relativistic equation, a wave function for the electron represented by a vector with four complex components, and which is a spinor of four-dimensional pre-Euclidean space–time.

Without knowing it, the physicists had just rediscovered the mathematical entities which had been created and studied by Élie Cartan [3] in 1913 when studying the linear representations of groups. We shall, in fact, see that the spinors which are useful to physicists are directly related to the rotation groups and to the unitary groups.

The spinors studied by Élie Cartan, moreover, are only one particular example of the spinors which can be defined in a very general way from certain axioms. Furthermore, the vector spaces which have spinors as elements are connected with the general theory of Clifford spaces, which were introduced by the mathematician W.K. Clifford in 1876.

On restricting ourselves to some particular cases of Clifford algebras, as has been done in this book, we can, however, present in a straightforward way the spinors which are useful in quantum physics without becoming lost in the vast field of these algebras.

1.1 Definition of a Spinor

1.1.1 Stereographic Projection

The study of the geometric properties of spinors allows us to understand better the relation between rotations in three-dimensional physical space and 'rotations' in the space of spinors. The use of the projection of the points of a sphere onto a plane by the method called stereographic projection allows us a simple introduction to spinors. Furthermore, we can then conceive how the transformations of the two-dimensional space of spinors stand in relation to the rotations in physical space.

Let us consider the coordinates (x, y, z) of a point P of a sphere with centre O and unit radius (Figure 1.1). We have

$$x^2 + y^2 + z^2 = 1. \tag{1.1.1}$$

Let us denote by N and S the points of intersection of the axis Oz with the sphere. The point S has coordinates $x = 0$, $y = 0$, $z = -1$. The stereographic projection P' of the point P is obtained by tracing the line SP which crosses the equatorial plane xOy at the point P' with coordinates (x', y', z'). The similar triangles $SP'O$ and SPQ give the relations

$$\begin{aligned} \frac{SP'}{SP} = \frac{x'}{x} = \frac{y'}{y} &= \frac{1}{1+z} \\ &= \frac{x' + iy'}{x + iy} = \frac{x' - iy'}{x - iy}. \end{aligned} \tag{1.1.2}$$

Let us set

$$\zeta = x' + iy', \qquad \zeta^* = x' - iy'.$$

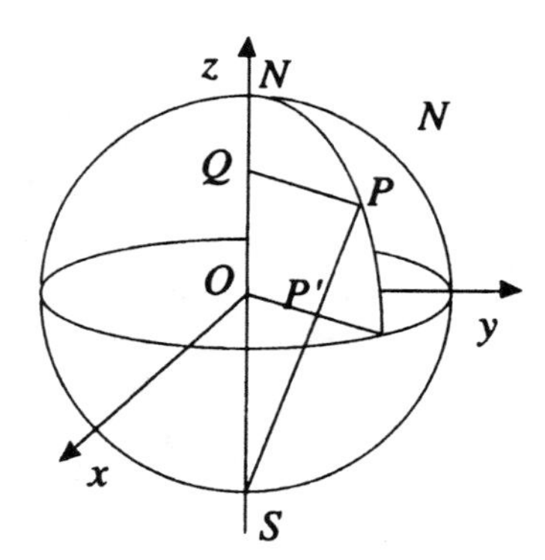

FIGURE 1.1. Stereographic projection.

Taking into account (1.1.1) this yields

$$\frac{1}{(1+z)^2} = \frac{\zeta\zeta^*}{x^2+y^2} = \frac{\zeta\zeta^*}{(1-z)(1+z)}. \tag{1.1.3}$$

Let us put the complex number ζ into the form

$$\zeta = \frac{\phi}{\psi}, \tag{1.1.4}$$

where ϕ and ψ are two complex numbers upon which we can always impose the unitarity condition

$$\psi\psi^* + \phi\phi^* = 1. \tag{1.1.5}$$

Relations (1.1.3), (1.1.4), and (1.1.5) give us the expression for z

$$z = \psi\psi^* - \phi\phi^*. \tag{1.1.6}$$

Relation (1.1.2) next gives us

$$x + iy = \zeta(1+z), \qquad x - iy = \zeta^*(1+z), \tag{1.1.7}$$

whence

$$x = \tfrac{1}{2}(\zeta + \zeta^*)(1+z), \qquad y = -\tfrac{1}{2}(\zeta - \zeta^*)(1+z). \tag{1.1.8}$$

Taking account of relations (1.1.4), (1.1.5), and (1.1.6) we obtain

$$x = \psi\phi^* + \psi^*\phi, \qquad y = i(\psi\phi^* - \psi^*\phi). \tag{1.1.9}$$

Thus at every point P sited on the sphere of unit radius we can bring into correspondence a pair (ψ, ϕ) of complex numbers satisfying the unitarity relation (1.1.5). The pair of complex numbers constitutes a UNITARY SPINOR.

We shall see that the spinors are able to constitute the elements of a two-dimensional vector space. The stereographic projection thus leads to representing certain vectors of the three-dimensional Euclidean space E_3 by elements of a two-dimensional complex vector space which is the vector space of spinors. Let us note that this representation is not unique. In fact, the arguments of ψ and ϕ are determined only to within a constant as a result of (1.1.4).

1.1.2 Vectors Associated with a Spinor

We are going to show that we can find two new vectors $\boldsymbol{X}_1$ and $\boldsymbol{X}_2$ of the Euclidean space E_3 which are associated the unitary spinor (ψ, ϕ) defined on the unit sphere. These vectors will be sought as mutually orthogonal and with unit norm, each being orthogonal to the vector $\boldsymbol{OP}$.

Let us write $\boldsymbol{X}_3 = \boldsymbol{OP}$ and (x_i, y_i, z_i), with $i = 1, 2, 3$, for the respective components of the vectors $\boldsymbol{X}_1, \boldsymbol{X}_2, \boldsymbol{X}_3$. The vectors sought are related by the vector product

$$\boldsymbol{X}_3 = \boldsymbol{X}_1 \times \boldsymbol{X}_2, \tag{1.1.10}$$

whence, taking account of the expression for the components of $\boldsymbol{OP}$ as functions of the components of the associated spinor, as well as the relation $\psi\psi^* + \phi\phi^* = 1$, we obtain

$$\begin{aligned} x_3 &= y_1 z_2 - y_2 z_1 \\ &= (\psi\psi^* + \phi\phi^*)(\psi\phi^* + \psi^*\phi), \\ y_3 &= z_1 x_2 - z_2 x_1 \\ &= i(\psi\psi^* + \phi\phi^*)(\psi\phi^* - \psi^*\phi), \\ z_3 &= x_1 y_2 - y_1 x_2 \\ &= (\psi\psi^* + \phi\phi^*)(\psi\psi^* - \phi\phi^*). \end{aligned} \tag{1.1.11}$$

On writing the mutual orthogonality of the vectors as well as the value of their norms we obtain six additional equations. However, as the orientation of the vectors $\boldsymbol{X}_1$ and $\boldsymbol{X}_2$ is not fixed there exists a certain indeterminacy in the values of their components. Let us choose values such that

$$\begin{aligned} x_1 + ix_2 &= \psi^2 - \phi^2, \\ y_1 + iy_2 &= i(\psi^2 + \phi^2), \\ z_1 + iz_2 &= -2\psi\phi. \end{aligned} \tag{1.1.12}$$

Taking the complex conjugates of the preceding relations, by addition of the components of $\boldsymbol{X}_1$ we obtain

$$\begin{aligned} x_1 &= \tfrac{1}{2}(\psi^2 - \psi^{*^2} - \phi^2 - \phi^{*2}), \\ y_1 &= \tfrac{1}{2}i(\psi^2 - \psi^{*2} + \phi^2 - \phi^{*2}), \\ z_1 &= -(\psi\phi + \psi^*\phi^*). \end{aligned} \tag{1.1.13}$$

By subtraction we obtain similarly the components of the vector $\boldsymbol{X}_2$

$$\begin{aligned} x_2 &= \tfrac{1}{2}i(-\psi^2 + \psi^{*2} + \phi^2 - \phi^{*2}), \\ y_2 &= \tfrac{1}{2}(\psi^2 + \psi^{*2} + \phi^2 + \phi^{*2}) \\ z_2 &= i(\psi\phi - \psi^*\phi^*). \end{aligned} \tag{1.1.14}$$

It is easily verified that these values again give (1.1.11) of the vector product. With every unitary spinor (ψ, ϕ) we can thus associate three vectors $\boldsymbol{X}_1, \boldsymbol{X}_2, \boldsymbol{X}_3$. We can verify quite straightforwardly that the vectors so calculated are mutually orthogonal and have unit norm.

The Unitary Vector $\boldsymbol{ON}$

By way of an example let us consider the vector $\boldsymbol{ON}$ (Figure 1.1) with components $(0, 0, 1)$. Equations (1.1.11) are satisfied by the associated spinor $(\psi, \phi) = (1, 0)$. Expressions (1.1.13) and (1.1.14) give us for its components $\boldsymbol{X}_1 = (1, 0, 0)$ and $\boldsymbol{X}_2 = (0, 1, 0)$. These are vectors with unit norm, taken, respectively, along the axes Ox and Oy.

1.1.3 The Definition of a Spinor

We are going to use the previous results to define a spinor, as Élie Cartan did, from the vectors $\boldsymbol{X}_1$ and $\boldsymbol{X}_2$. This definition is clearly equivalent to that given previously by the stereographic projection using only one vector. However, using two vectors allows us to define an oriented plane, and thus a direction of rotation, by specifying the *order* in which we pick these vectors.

Consider two vectors $\boldsymbol{X}_1 = (x_1, y_1, z_1)$ and $\boldsymbol{X}_2 = (x_2, y_2, z_2)$ of the Euclidean space E_3 with the same origin. which are mutually orthogonal and have equal norms. These vectors define a plane, and if we consider these vectors as being *ordered* this order defines a direction of rotation. Let us make the choice of the order of $\boldsymbol{X}_1, \boldsymbol{X}_2$ corresponding to the indices. We can consider that a plane has been defined which is oriented in an intrinsic manner.

Since these vectors are orthogonal and have the same norm we can write

$$\begin{aligned} \|\boldsymbol{X}_1\|^2 = \|\boldsymbol{X}_2\|^2 &= x_1^2 + y_1^2 + z_1^2 \\ &= x_2^2 + y_2^2 + z_2^2, \end{aligned} \tag{1.1.15}$$

$$\begin{aligned} \boldsymbol{X}_1 \cdot \boldsymbol{X}_2 &= x_1 x_2 + y_1 y_2 + z_1 z_2 \\ &= 0. \end{aligned} \tag{1.1.16}$$

Isotropic Vectors

In order to introduce algebraically a representation of the order in which the vectors $\boldsymbol{X}_1$ and $\boldsymbol{X}_2$ have been considered, it is convenient to multiply the components of the second vector by the imaginary number i. We can thus form the three complex numbers

$$x = x_1 + ix_2, \qquad y = y_1 + iy_2, \qquad z = z_1 + iz_2. \tag{1.1.17}$$

These three complex numbers are able to form the components of a vector

$$\boldsymbol{Z} = (x, y, z). \tag{1.1.18}$$

Consequently, from the choice of the vectors $\boldsymbol{X}_1$ and $\boldsymbol{X}_2$ the three complex numbers x, y, and z are not independent. In fact, we have

$$\begin{aligned} x^2 + y^2 + z^2 &= (x_1)^2 + (y_1)^2 + (z_1)^2 - [(x_2)^2 + (y_2)^2 + (z_2)^2] \\ &\quad + 2i(x_1 x_2 + y_1 y_2 + z_1 z_2). \end{aligned} \tag{1.1.19}$$

The previous relations (1.1.15) and (1.1.15) give us

$$\begin{aligned} \boldsymbol{Z}\cdot\boldsymbol{Z} = \|\boldsymbol{Z}\|^2 &= x^2 + y^2 + z^2 \\ &= 0. \end{aligned} \tag{1.1.20}$$

Such a vector $\boldsymbol{Z}$ of zero norm is called an ISOTROPIC VECTOR. It is orthogonal to itself.

Spinors

Relation (1.1.20) between the three complex numbers x, y, z allows us to express them by means of just two complex numbers. This relation can be put into the form

$$\begin{aligned} z^2 &= -(x^2 + y^2) \\ &= -(x + iy)(x - iy). \end{aligned} \tag{1.1.21}$$

If we set

$$x + iy = -2\phi^2, \qquad x - iy = 2\psi^2, \tag{1.1.22}$$

these two complex numbers ψ and ϕ allow us to calculate x, y, z. In fact, we have

$$x = \psi^2 - \phi^2, \qquad y = i(\psi^2 - \phi^2), \qquad z = \pm 2\psi\phi. \tag{1.1.23}$$

One of the signs $+$ or $-$ can be chosen arbitrarily for the value of z. To preserve Élie Cartan's notation the negative sign will be chosen. Thus we have

$$\begin{aligned} x &= x_1 + ix_2 = \psi^2 - \phi^2, \\ y &= y_1 + iy_2 = i(\psi^2 + \phi^2), \\ z &= z_1 + iz_2 = -2\psi\phi. \end{aligned} \tag{1.1.24}$$

If two orthogonal vectors $\boldsymbol{X}_1$ and $\boldsymbol{X}_2$ with the same norm are given, the two complex numbers ψ and ϕ are determined by relations (1.1.24). Conversely, the decomposition of the expressions $\psi^2 - \phi^2$, $i(\psi^2 + \phi^2)$, and $-2\psi\phi$ into their real and imaginary parts gives the respective components of $\boldsymbol{X}_1$ and $\boldsymbol{X}_2$.

The set of two complex numbers ψ and ϕ thus forms a representation of the two vectors $\boldsymbol{X}_1$ and $\boldsymbol{X}_2$ as well as of the order chosen for these two vectors. The pairs of number (ψ, ϕ) related to the vectors $\boldsymbol{X}_1$ and $\boldsymbol{X}_2$ by relations (1.1.24) constitutes a *spinor*.

The Unitary Spinor

If we consider the vectors $\boldsymbol{X}_1$ and $\boldsymbol{X}_2$ with unit norm, the associated spinor (ψ, ϕ) satisfies the relation

$$\begin{aligned} \|\boldsymbol{X}_1\| = \|\boldsymbol{X}_2\| &= 1 \\ &= \psi\psi^* + \phi\phi^*. \end{aligned} \tag{1.1.25}$$

Such a spinor is called a UNITARY SPINOR. The complex numbers forming a unitary spinor may be written in the form

$$\psi = \cos\alpha e^{i\beta}, \qquad \phi = \sin\alpha e^{i\gamma}. \tag{1.1.26}$$

A unitary spinor thus depends upon three angular components here denoted α, β, γ. Similarly, an arbitrary spinor can be put into the form

$$\psi = \rho\cos\alpha e^{i\beta}, \qquad \phi = \rho\sin\alpha e^{i\gamma}. \tag{1.1.27}$$

Associated Vectors

Conversely, if two complex numbers (ψ, ϕ) are given we can determine two vectors $\boldsymbol{X}_1, \boldsymbol{X}_2$ respectively satisfying the relations (1.1.13) and (1.1.14). A third vector $\boldsymbol{X}_3$ can also be determined by the vector product $\boldsymbol{X}_3 = \boldsymbol{X}_1 \times \boldsymbol{X}_2$.

1.2 Geometrical Properties

We are going to study the transformations of vectors of E_3 associated with a spinor in order to deduce from them the corresponding properties of transformations of spinors. Since rotations may be expressed as the product of two plane symmetries we begin with the study of the latter.

1.2.1 Plane Symmetries

Plane Symmetry of a Vector

Under a symmetry with respect to a plane $\mathcal{P}$ an arbitrary vector $\boldsymbol{OM}$ is transformed into a vector $\boldsymbol{OM}'$. Let us determine a matrix $\mathcal{S}$ which represents this symmetry about this plane. Let us choose the origin O of a reference trihedron on the plane $\mathcal{P}$. Let $\boldsymbol{OA}$ be a unitary vector described by a hemi-normal to the plane $\mathcal{P}$ and let H be the foot of the perpendicular dropped from a point M of the space onto the plane $\mathcal{P}$ (Figure 1.2).

Let M' be the point symmetric with M with respect to $\mathcal{P}$. We have

$$\begin{aligned} \boldsymbol{OM}' &= \boldsymbol{OM} + \boldsymbol{MM}' \\ &= \boldsymbol{OM} + 2\boldsymbol{MH} \\ &= \boldsymbol{OM} - 2\boldsymbol{OA}(\boldsymbol{OM}\cdot\boldsymbol{OA}). \end{aligned} \tag{1.2.1}$$

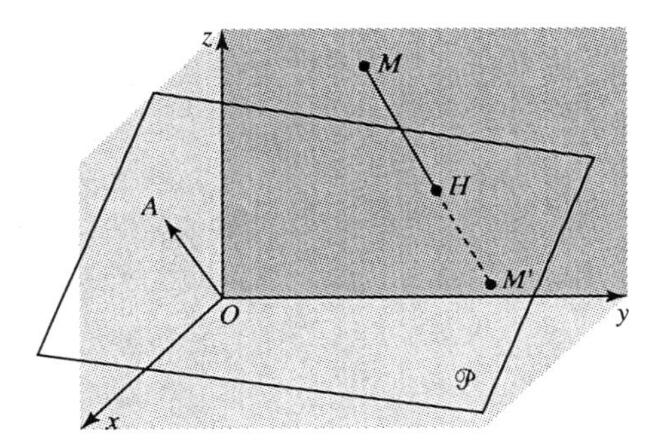

FIGURE 1.2. Plane symmetry of a vector.

Let a_1, a_2, a_3 be the Cartesian coordinates of $\boldsymbol{OA}$. x, y, z and x', y', z' are, respectively, the components of the vectors $\boldsymbol{OM}$ and $\boldsymbol{OM}'$. The relation (1.2.1) gives us the linear relations

$$\begin{aligned} x' &= x - 2a_1(a_1x + a_2y + a_3z), \\ y' &= y - 2a_2(a_1x + a_2y + a_3z), \\ z' &= z - 2a_3(a_1x + a_2y + a_3z). \end{aligned} \tag{1.2.2}$$

The matrix $\mathcal{S}$ which takes the vector $\boldsymbol{X} = (x, y, z)$ into the vector $\boldsymbol{X}' = (x', y', z')$ thus has the expression

$$\mathcal{S} = \begin{bmatrix} 1 - 2(a_1)^2 & -2a_1a_2 & -2a_1a_3 \\ -2a_2a_1 & 1 - 2(a_2)^2 & -2a_2a_3 \\ -2a_3a_1 & -2a_3a_2 & 1 - 2(a_3)^2 \end{bmatrix}. \tag{1.2.3}$$

We can verify that $\mathcal{S}$ is a unitary matrix such that $4|\mathcal{S}| = -1$ by using the fact

$$(a_1)^2 + (a_2)^2 + (a_3)^2 = 1.$$

Plane Symmetry of Two Vectors

Let us consider for the moment two mutually orthogonal and unitary vectors $\boldsymbol{X}_1$ and $\boldsymbol{X}_2$ defining a unitary spinor (ψ, ϕ). A symmetry with respect to a plane $\mathcal{P}$ transforms the vectors $\boldsymbol{X}_1$ and $\boldsymbol{X}_2$ into vectors $\boldsymbol{X}_1'$ and $\boldsymbol{X}_2'$ which has the spinor (ψ', ϕ') associated with them. What is the expression for the transformation which takes the spinor (ψ, ϕ) into the spinor (ψ', ϕ')?

Let us write $\boldsymbol{A} = (a_1, a_2, a_3)$ for the unitary vector defining one of the hemi-normals to the plane $\mathcal{P}$ in three-dimensional space. We are going to show that the following transformation of the spinor (ψ, ϕ) into (ψ', ϕ')

$$\begin{bmatrix} \psi' \\ \phi' \end{bmatrix} = \begin{bmatrix} a_3 & a_1 - ia_2 \\ a_1 + ia_2 & -a_3 \end{bmatrix} \begin{bmatrix} \psi \\ \phi \end{bmatrix} = A \begin{bmatrix} \psi \\ \phi \end{bmatrix} \tag{1.2.4}$$

exactly transforms the vectors $(\boldsymbol{X}_1, \boldsymbol{X}_2)$ into the vectors $(\boldsymbol{X}_1{}', \boldsymbol{X}_2{}')$, these vectors being respectively deduced from each other by a plane symmetry. The matrix A certainly represents the transformation sought.

For that let us calculate the components of the vector $\boldsymbol{X}' = \boldsymbol{X}_1{}' + i\boldsymbol{X}_2{}'$ as a function of the components of $\boldsymbol{X} = \boldsymbol{X}_1 + i\boldsymbol{X}_2$. Relation (1.2.4) gives us

$$\psi' = a_3\psi + (a_1 - ia_2)\phi, \qquad \phi' = (a_1 + ia_2)\psi - a_3\phi. \tag{1.2.5}$$

Let us write the components of $\boldsymbol{X}$ as (x, y, z) and those of $\boldsymbol{X}'$ as $(x', y', z',)$. According to the defining formulas (1.1.24) and taking (1.2.5) into account we have

$$\begin{aligned} x' = \psi'^2 - \phi'^2 &= [-(a_1)^2 + (a_2)^2 + (a_3)^2](\psi^2 - \phi^2) \\ &\quad -2a_1a_2i(\psi^2 + \phi^2) - 2a_1a_3(-2\psi\phi), \\ y' = i(\psi'^2 + \phi'^2) &= -2a_2a_1(\psi^2 - \phi^2) - 2a_2a_3(-2\psi\phi) \\ &\quad +[(a_1)^2 - (a_2)^2 + (a_3)^2]i(\psi^2 + \phi^2), \\ z' = -2\psi'\phi' &= -2a_3a_1(\psi^2 - \phi^2) - 2a_2a_3i(\psi^2 + \phi^2) \\ &\quad +[(a_1)^2 + (a_2)^2 - (a_3)^2](-2\psi\phi). \end{aligned} \tag{1.2.6}$$

Consequently from the relation

$$a_1^2 + a_2^2 + a_3^2 = 1$$

we obtain

$$\begin{aligned} x' &= [1 - 2(a_1)^2]x - 2a_1a_2y - 2a_1a_3z, \\ y' &= -2a_1a_2z + [1 - 2(a_2)^2]y - 2a_2a_3z, \\ z' &= -2a_3a_1x - 2a_3a_2y + [1 - 2(a_3)^2]z. \end{aligned} \tag{1.2.7}$$

The separation of the components x', y', z' and x, y, z into their real and imaginary parts gives the components of $\boldsymbol{X}' = (x_1', y_1', z_1')$ as functions of those of $\boldsymbol{X}_2 = (x_1, y_1, z_1)$, and similarly for $\boldsymbol{X}_2{}'$ as a function of $\boldsymbol{X}_2$. For $\boldsymbol{X}_1$ and $i\boldsymbol{X}_2$ we obtain the same transformation matrix $\mathcal{S}$ given by (1.2.3) which appears in expression (1.2.7).

The matrix A thus generates the transformation of a spinor (ψ, ϕ) into a spinor (ψ', ϕ') such that the associated vectors $(\boldsymbol{X}_1, \boldsymbol{X}_2)$ are respectively deduced from $(\boldsymbol{X}_1{}', \boldsymbol{X}_2{}')$ by a plane symmetry.

1.2.2 *Rotations*

Figure 1.3 shows that the product of symmetries about the lines Δ and Δ' is equivalent to a rotation through an angle $2(\alpha + \beta)$ about the point of intersection O when the angle between the two lines is equal to $(\alpha + \beta)$. Similarly, for two oriented planes $\mathcal{P}$ and $\mathcal{Q}$ having a line of intersection $\mathcal{L}$

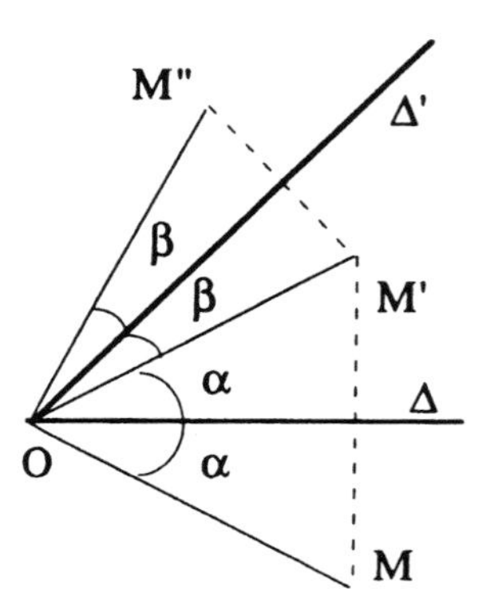

FIGURE 1.3. The product of symmetries about two lines.

and their dihedron formed by these planes has an angle equal to $\frac{1}{2}\theta$, the product of the symmetries with respect to these planes is equivalent to a rotation through an angle θ about the line of intersection $\mathcal{L}$.

The rotations in the Euclidean vector space are going to be studied by considering the product of two plane symmetries. These symmetries are oriented in a direction of rotation defined in each plane by pairs of vectors $(\boldsymbol{X}_1, \boldsymbol{X}_2)$. The hemi-normals $\boldsymbol{X}_3$ associated with each plane are chosen in such a way that the vectors $(\boldsymbol{X}_1, \boldsymbol{X}_2, \boldsymbol{X}_3)$ define a straightforward trihedron. The choice of the same type of trihedron for each of the triplets of vectors with respect to a plane is necessary in order that the product of the two plane symmetries provide a rotation and not the product of a rotation with a reflection.

Let us consider two oriented planes $\mathcal{P}$ and $\mathcal{Q}$ forming a dihedron, and let us write $\boldsymbol{A} = (a_1, a_2, a_3)$ and $\boldsymbol{B} = (b_1, b_2, b_3)$ for unitary vectors along the respective hemi-normals to the planes $\mathcal{P}$ and $\mathcal{Q}$. Let us write the angle of the dihedron as $\frac{1}{2}\theta$ and consequently it is the angle between the vectors $\boldsymbol{A}$ and $\boldsymbol{B}$. Let $\boldsymbol{L}$ be the unitary vector made by the line $\mathcal{L}$ resulting from the intersection of the planes $\mathcal{P}$ and $\mathcal{Q}$ (Figure 1.4) and such that

$$\boldsymbol{A} \times \boldsymbol{B} = \boldsymbol{L} \sin \tfrac{1}{2}\theta. \tag{1.2.8}$$

The vectors $\boldsymbol{A}, \boldsymbol{B}, \boldsymbol{L}$ form a simple trihedron. Let us write L_1, L_2, L_3 for the components of the vectors $\boldsymbol{L}$. We have

$$(L_1)^2 + (L_2)^2 + (L_3)^2 = 1. \tag{1.2.9}$$

Relation (1.2.8) gives us for the components of $\boldsymbol{L}$

$$\begin{aligned} L_1 \sin \tfrac{1}{2}\theta &= a_2 b_3 - a_3 b_2, \\ L_2 \sin \tfrac{1}{2}\theta &= a_3 b_1 - a_1 b_3, \\ L_3 \sin \tfrac{1}{2}\theta &= a_1 b_2 - a_2 b_1. \end{aligned} \tag{1.2.10}$$

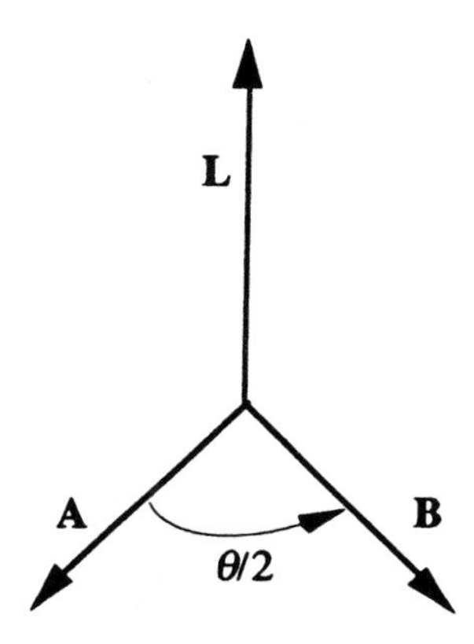

FIGURE 1.4. The angle of the dihedron.

On the other hand the scalar product $\boldsymbol{A}\cdot\boldsymbol{B}$ may be written

$$\boldsymbol{A}\cdot\boldsymbol{B} = a_1 b_1 + a_2 b_2 + a_3 b_3 = \cos \tfrac{1}{2}\theta. \tag{1.2.11}$$

The Rotation Matrix

Let us calculate the matrix of the transformation which, when acting on a spinor (ψ, ϕ) yields a spinor (ψ', ϕ'), corresponds to a rotation of the associated vectors in the Euclidean space. This rotation through an angle θ (double the angle of the dihedron) determined by the product of two plane symmetries is represented in the space of spinors by the product of matrices of the type (1.2.4), that is,

$$\begin{aligned} R(\boldsymbol{L}, \tfrac{1}{2}\theta) &= BA \\ &= \begin{bmatrix} b_3 & b_1 - ib_2 \\ b_1 + ib_2 & -b_3 \end{bmatrix} \begin{bmatrix} a_3 & a_1 - ia_2 \\ a_1 + ia_2 & -a_3 \end{bmatrix}. \end{aligned} \tag{1.2.12}$$

Expanding the product of the matrices and taking (1.2.10) and (1.2.11) into account we obtain

$$R(\boldsymbol{L}, \tfrac{1}{2}\theta) = \begin{bmatrix} \cos\frac{1}{2}\theta - iL_3 \sin\frac{1}{2}\theta & -(iL_1 - L_2)\sin\frac{1}{2}\theta \\ (-iL_1 + L_2)\sin\frac{1}{2}\theta & \cos\frac{1}{2}\theta + iL_3\sin\frac{1}{2}\theta \end{bmatrix}. \tag{1.2.13}$$

The determinant of this matrix is equal to unity. It will be said that the transformation of the spinor (ψ, ϕ) into a spinor (ψ', ϕ') is a ROTATION *in the space of spinors*, and this rotation is defined by the matrix $R(\boldsymbol{L}, \frac{1}{2}\theta)$. This matrix has the form

$$R = \begin{bmatrix} a & b \\ -b^* & a^* \end{bmatrix}. \tag{1.2.14}$$

This is a unitary matrix such that $RR^{-1} = I$, where I is the unit matrix. If the angle of rotation is changed from $\frac{1}{2}\theta$ into $-\frac{1}{2}\theta$ we obtain the matrix R^{-1} which corresponds to the inverse rotation of that corresponding to R. The parameters $a, b, -b^*, a^*$ are called the CAYLEY–KLEIN PARAMETERS *of the rotation.*

A New Definition of Spinors

Given the importance of rotations in the use of spinors in physics, for the moment we are going to characterize a spinor by the transformation property (1.2.13) corresponding to a rotation in three-dimensional space.

DEFINITION **1.1.** *We call a spinor every pair* (ψ, ϕ) *of complex numbers which under a rotation through an angle* θ *in three-dimensional space is transformed into a pair* (ψ', ϕ') *by the matrix* $R(\boldsymbol{L}, \frac{1}{2}\theta)$ *given by* (1.2.13). ■

1.2.3 The Olinde–Rodrigues Parameters

Let two unitary spinors (ψ, ϕ) and (ψ', ϕ') be given. It is proposed to determine the rotation matrix $R(\boldsymbol{L}, \frac{1}{2}\theta)$ which allows us to go from one spinor to the other, that is to say, to calculate the four elements of the matrix $R(\boldsymbol{L}, \frac{1}{2}\theta)$ as a function of the given four complex numbers ψ, ϕ, ψ', ϕ'. Since the spinors are unitary we have

$$\psi\psi^* + \phi\phi^* = \psi\psi'^* + \phi\phi'^* = 1. \tag{1.2.15}$$

By hypothesis we have the following relation between the spinors

$$\begin{bmatrix} \psi' \\ \phi' \end{bmatrix} = R(\boldsymbol{L}, \tfrac{1}{2}\theta) \begin{bmatrix} \psi \\ \phi \end{bmatrix} \tag{1.2.16}$$

which gives us the following equations

$$\begin{aligned} \psi' &= [\cos\tfrac{1}{2}\theta - iL_3 \sin\tfrac{1}{2}\theta]\psi - [(iL_1 + L_2)\sin\tfrac{1}{2}\theta]\phi, \\ \phi' &= [(-iL_1 + L_2)\sin\tfrac{1}{2}\theta]\psi + [\cos\tfrac{1}{2}\theta + iL_3 \sin\tfrac{1}{2}\theta]\phi. \end{aligned} \tag{1.2.17}$$

The complex conjugates of expressions (1.2.17) give us

$$\begin{aligned} \psi'^* &= [\cos\tfrac{1}{2}\theta + iL_3 \sin\tfrac{1}{2}\theta]\psi^* - [(-iL_1 + L_2)\sin\tfrac{1}{2}\theta]\phi^*, \\ \phi'^* &= [(-iL_1 + L_2)\sin\tfrac{1}{2}\theta]\psi^* + [\cos\tfrac{1}{2}\theta - iL_3 \sin\tfrac{1}{2}\theta]\phi^*. \end{aligned} \tag{1.2.18}$$

Multiplying the first equation of (1.2.17) by ψ^* and the second of (1.2.18) by ϕ, and taking account of the unitarity relation (1.2.15), we obtain

$$\cos\tfrac{1}{2}\theta - iL_3 \sin\tfrac{1}{2}\theta = \psi'\psi^* + \phi'^*\phi'. \tag{1.2.19}$$

Multiplying the first equation of (1.2.17) by $-\phi^*$ and the second of (1.2.18) by ψ we obtain, similarly,

$$(iL_1 + L_2)\sin\tfrac{1}{2}\theta = \phi'^*\psi - \psi'\phi^*. \tag{1.2.20}$$

These two last equations allow us to obtain the following parameters

$$\cos\tfrac{1}{2}\theta, \quad L_1\sin\tfrac{1}{2}\theta, \quad L_2\sin\tfrac{1}{2}\theta, \quad L_3\sin\tfrac{1}{2}\theta, \tag{1.2.21}$$

which are called the OLINDE–RODRIGUES PARAMETERS. These parameters give, by combination, the Cayley–Klein parameters which determine the rotation matrix $R(\boldsymbol{L}, \frac{1}{2}\theta)$. As the vector $\boldsymbol{L}$ is unitary its components satisfy the relation

$$(L_1)^2 + (L_2)^2 + (L_3)^2 = 1, \tag{1.2.22}$$

and as a result the Olinde–Rodrigues parameters are such that

$$\cos^2\tfrac{1}{2}\theta + \sin^2\tfrac{1}{2}\theta((L_1)^2 + (L_2)^2 + (L_3)^2) = 1. \tag{1.2.23}$$

The solution of (1.2.19) and (1.2.20) with respect to the Olinde–Rodrigues parameters gives us

$$\begin{aligned}
\cos\tfrac{1}{2}\theta &= \tfrac{1}{2}(\psi'\psi^* + \psi'^*\psi + \phi'^*\phi + \phi'\phi^*),\\
L_1\sin\tfrac{1}{2}\theta &= -\tfrac{1}{2}i(\phi'^* - \phi'\psi^* - \psi'\phi^* + \psi'^*\phi),\\
L_2\sin\tfrac{1}{2}\theta &= \tfrac{1}{2}(\phi'^* + \phi'\psi^* - \psi'\phi^* - \psi'^*\phi),\\
L_3\sin\tfrac{1}{2}\theta &= -\tfrac{1}{2}i(\psi'^* - \psi'\psi^* + \phi'\phi^* - \phi'^*\phi).
\end{aligned} \tag{1.2.24}$$

The rotation matrix $R(\boldsymbol{L}, \frac{1}{2}\theta)$ is thus completely determined by the given spinors (ψ, ϕ) and (ψ', ϕ').

1.2.4 Rotations Defined in Terms of the Euler Angles

Let us consider a fixed frame of reference $Oxyz$ of which we are going to study the rotation about an axis passing through the point O. Initially the moving trihedron coincides with the fixed trihedron $Oxyz$.

Every rotation about an axis passing through the point O can be decomposed into three successive rotations defined in the following way (Figure 1.5)

(1) First, a rotation through an angle α about Oz, which in the plane xOy takes the axis Ox' into Ox_1 and the axis Oy' into Oy_1.

(2) Second, a rotation through an angle β about the axis Ox_1, which in the plane y_1Oz takes the axis Oz' into Oz_2 and the axis Oy_1 into Oy_2.

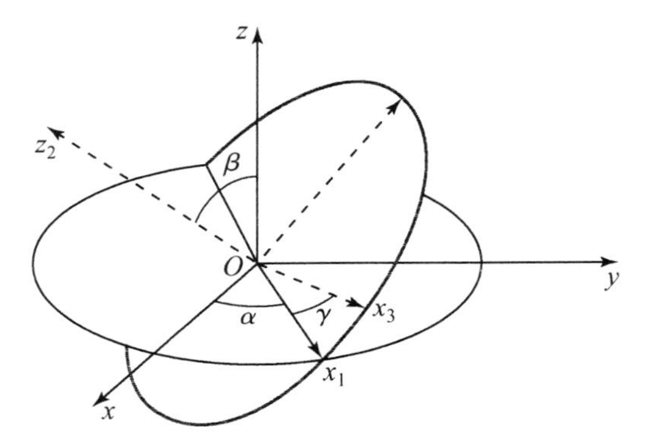

FIGURE 1.5. The decomposition of a rotation into three successive rotations defined by the Euler angles.

(3) Third, a rotation through an angle γ about the axis Oz_2, which in the plane x_1Oy_2 takes the axis Ox_1 into Ox_3 and the axis Oy_2 into Oy_3.

The angles α, β, γ are called the EULER ANGLES. We are going to determine the rotation matrix $R(\boldsymbol{L}, \frac{1}{2}\theta)$ as a function of three rotations given by the Euler angles. To do so it suffices to determine each of the matrices corresponding to one of the successive rotations and then to form the product of the matrices.

The first rotation introduces the unitary vector $\boldsymbol{L}_1 = (0, 0, 1)$ and the angle of rotation α. The rotation matrix $R(\boldsymbol{L}, \frac{1}{2}\theta)$ in this spinor space is obtained by setting $L_1 = 0$, $L_2 = 0$, and $L_3 = 1$ in the matrix and considering an angle of $\frac{1}{2}\alpha$, whence

$$\begin{aligned} R(Oz, \tfrac{1}{2}\alpha) &= \begin{bmatrix} \cos\frac{1}{2}\alpha - i\sin\frac{1}{2}\alpha & 0 \\ 0 & \cos\frac{1}{2}\alpha + i\sin\frac{1}{2}\alpha \end{bmatrix} \\ &= \begin{bmatrix} e^{-i\alpha/2} & 0 \\ 0 & e^{i\alpha/2} \end{bmatrix}. \end{aligned} \tag{1.2.25}$$

The second rotation is carried out about the axis Ox_1, along the unitary vector $\boldsymbol{L}_2 = (\cos\alpha, \sin\alpha, 0)$ through a rotation of an angle of β, whence the matrix

$$\begin{aligned} R(Ox_1, \tfrac{1}{2}\beta) &= \begin{bmatrix} \cos\frac{1}{2}\beta & -(i\cos\alpha + \sin\alpha)\sin\frac{1}{2}\beta \\ -(i\cos\alpha - \sin\alpha)\sin\frac{1}{2}\beta & \cos\frac{1}{2}\beta \end{bmatrix} \\ &= \begin{bmatrix} \cos\frac{1}{2}\beta & -ie^{-i\alpha}\sin\frac{1}{2}\beta \\ -ie^{i\alpha}\sin\frac{1}{2}\beta & \cos\frac{1}{2}\beta \end{bmatrix}. \end{aligned} \tag{1.2.26}$$

The third rotation through an angle γ is made about the axis Oz_2 along the unitary vector $\boldsymbol{L}_3 = (\sin\alpha\sin\beta, -\cos\alpha\sin\beta, \cos\beta)$, whence the matrix

$$R(Oz_2, \tfrac{1}{2}\gamma) = \begin{bmatrix} \cos\frac{1}{2}\gamma - i\cos\beta\sin\frac{1}{2}\gamma & e^{-i\alpha}\sin\beta\sin\frac{1}{2}\gamma \\ -e^{i\alpha}\sin\beta\sin\frac{1}{2}\gamma & \cos\frac{1}{2}\gamma + i\cos\beta\sin\frac{1}{2}\gamma \end{bmatrix}. \quad (1.2.27)$$

Finally, the complete rotation $R(\alpha, \beta, \gamma)$ which results from the successive application of these three rotations is represented for spinors by the matrix product

$$R(\alpha, \beta, \gamma) = R(Oz_2, \tfrac{1}{2}\gamma)R(Ox_1, \tfrac{1}{2}\beta)R(OZ, \tfrac{1}{2}\alpha). \quad (1.2.28)$$

The product of the matrices gives us

$$R(\alpha, \beta, \gamma) = \begin{bmatrix} e^{-i(\alpha+\gamma)/2}\cos\frac{1}{2}\beta & -ie^{-i(\alpha-\gamma)/2}\sin\frac{1}{2}\beta \\ -ie^{i(\alpha-\gamma)/2}\sin\frac{1}{2}\beta & e^{i(\alpha+\gamma)/2}\cos\frac{1}{2}\beta \end{bmatrix}. \quad (1.2.29)$$

Every spinor transforms according to the preceding matrix under a rotation defined by the angles (α, β, γ). Let us consider, for example, the unitary spinor $(1, 0)$ corresponding to the unitary vectors $\boldsymbol{X}_1, \boldsymbol{X}_2$ directed along the respective axes Ox and Oy. A rotation $R(\alpha, \beta, \gamma)$ transforms this spinor into another spinor (ψ', ϕ') such that

$$\psi' = e^{-i(\alpha+\gamma)/2}\cos\tfrac{1}{2}\beta, \qquad \phi' = -ie^{i(\alpha-\gamma)/2}\sin\tfrac{1}{2}\beta. \quad (1.2.30)$$

We obtain a spinor corresponding to the general form given by expressions (1.1.26).

REMARK **1.1.** *The Euler angles are not always defined in the above way. We have taken as the successive axes of rotation Oz, Ox_1, and Oz_2. If we take Oz, Oy_1, and Oz_2 as the axes of rotation, always keeping α, β, γ as the angles of rotation, we obtain an unchanged value for the terms a and a^* (the notation of* (1.2.14)*) of the matrix $R(\alpha, \beta, \gamma)$ which are located on the principal diagonal, but we obtain*

$$b = -\exp[-\tfrac{1}{2}i(\alpha - \gamma)]\sin\tfrac{1}{2}\beta. \qquad \blacksquare$$

1.3 Infinitesimal Properties of Rotations

1.3.1 The Infinitesimal Rotation Matrix

The rotation matrix $R(\boldsymbol{L}, \frac{1}{2}\theta)$ given by expression (1.2.13) can be written in the form of a limited expansion of infinitesimal rotations $\theta/2 = \varepsilon/2$. We have

$$R(\boldsymbol{L}, \tfrac{1}{2}\theta) = \begin{bmatrix} \cos\frac{1}{2}\varepsilon - iL_3\sin\frac{1}{2}\varepsilon & -(iL_1 + L_2)\sin\frac{1}{2}\varepsilon \\ (-iL_1 + L_2)\sin\frac{1}{2}\varepsilon & \cos\frac{1}{2}\varepsilon + iL_2\sin\frac{1}{2}\varepsilon \end{bmatrix}. \quad (1.3.1)$$

Let us write the functions $\cos\frac{1}{2}\varepsilon$ and $\sin\frac{1}{2}\varepsilon$ in the form of their restricted expansion. This gives

$$\cos\tfrac{1}{2}\varepsilon = 1 - \frac{(\frac{1}{2}\varepsilon)^2}{2} + \cdots, \qquad \sin\tfrac{1}{2}\varepsilon = \frac{\varepsilon}{2} - \frac{(\frac{1}{2}\varepsilon)^3}{6} + \cdots. \tag{1.3.2}$$

Using only the first-order terms the rotation matrix may be written

$$R(\boldsymbol{L}, \tfrac{1}{2}\theta) = \begin{bmatrix} 1 - iL_3\frac{1}{2}\varepsilon & -(iL_1 + L_2)\frac{1}{2}\varepsilon \\ (-iL_1 + L_2)\frac{1}{2}\varepsilon & 1 + iL_3\frac{1}{2}\varepsilon \end{bmatrix}. \tag{1.3.3}$$

This matrix is the restricted expansion of the rotation matrix in the neighborhood of the identity matrix, the latter corresponding to the zero rotation. The matrix (1.3.2) can be written

$$\begin{aligned} R(\boldsymbol{L}, \tfrac{1}{2}\theta) &= \begin{bmatrix} 1 & 0 \\ 0 & 1 \end{bmatrix} + \frac{\varepsilon}{2}\begin{bmatrix} -iL_3 & -(iL_1 + L_2) \\ (-iL_1 + L_2) & iL_3 \end{bmatrix} \\ &= \sigma_0 + \varepsilon X(\boldsymbol{L}). \end{aligned} \tag{1.3.4}$$

The matrix σ_0 is the unit matrix of order 2 and $X(\boldsymbol{L})$ is called the INFINITESIMAL ROTATION MATRIX. If we write $R_{ij}(\boldsymbol{L}, \frac{1}{2}\theta)$ for the elements of the matrix $R(\boldsymbol{L}, \frac{1}{2}\theta)$ given by (1.3.1), then the elements $X_{ij}(\boldsymbol{L})$ of the matrix $X(\boldsymbol{L})$ are given by

$$X_{ij}(\boldsymbol{L}) = \left[\frac{\partial R_{ij}(\boldsymbol{L}, \frac{1}{2}\varepsilon)}{\partial\varepsilon}\right]_{\varepsilon=0}, \tag{1.3.5}$$

and we have

$$R_{ij}(\boldsymbol{L}, \tfrac{1}{2}\varepsilon) = R_{ij}(\boldsymbol{L}, 0) + \varepsilon X_{ij}(\boldsymbol{L}),$$

the quantities $R_{ij}(\boldsymbol{L}, 0)$ being the elements of the unit matrix.

1.3.2 The Pauli Matrices

The matrix $R(\boldsymbol{L}, \frac{1}{2}\varepsilon)$ corresponds to a rotation in three-dimensional space through an angle ε about an axis lying along the unitary vector $\boldsymbol{L}$. Let us consider, for the moment, particular rotations through the same angle about each of the axes of the reference frame $Oxyz$.

For a rotation about the axis Ox the vector $\boldsymbol{L}$ has as its components

$$L_1 = 1, \qquad L_2 = 0, \qquad L_3 = 0, \tag{1.3.6}$$

whence, by (1.3.4), the expression for the infinitesimal matrix

$$\begin{aligned} X(Ox) &= \frac{1}{2}\begin{bmatrix} 0 & -i \\ -i & 0 \end{bmatrix} \\ &= -\frac{i}{2}\begin{bmatrix} 0 & 1 \\ 1 & 0 \end{bmatrix}. \end{aligned} \tag{1.3.7}$$

A rotation about the axis Oy corresponds to a vector $\boldsymbol{L}$ with components

$$L_1 = 0, \qquad L_2 = 0, \qquad L_3 = 0, \tag{1.3.8}$$

whence the infinitesimal matrix

$$\begin{aligned} X(Oy) &= \frac{1}{2}\begin{bmatrix} 0 & -1 \\ 1 & 0 \end{bmatrix} \\ &= -\frac{i}{2}\begin{bmatrix} 0 & -i \\ i & 0 \end{bmatrix}. \end{aligned} \tag{1.3.9}$$

A rotation about O corresponds to a vector $\boldsymbol{L}$ with components

$$L_1 = 0, \qquad L_2 = 0, \qquad L_3 = 1, \tag{1.3.10}$$

and for the infinitesimal matrix we obtain

$$\begin{aligned} X(Oz) &= \frac{1}{2}\begin{bmatrix} -i & 0 \\ 0 & i \end{bmatrix} \\ &= -\frac{i}{2}\begin{bmatrix} 1 & 0 \\ 0 & -1 \end{bmatrix}. \end{aligned} \tag{1.3.11}$$

The infinitesimal matrices (1.3.7), (1.3.9), and (1.3.11) bring into evidence the following matrices

$$\begin{aligned} \sigma_1 &= \begin{bmatrix} 0 & 1 \\ 1 & 0 \end{bmatrix}, \\ \sigma_2 &= \begin{bmatrix} 0 & -i \\ i & 0 \end{bmatrix}, \\ \sigma_3 &= \begin{bmatrix} 1 & 0 \\ 0 & -1 \end{bmatrix}. \end{aligned} \tag{1.3.12}$$

These matrices are called the PAULI MATRICES and they play a fundamental role in quantum mechanics.

The Expression for the Rotation Matrix in Terms of Pauli Matrices

The restricted expansion (1.3.4) of the rotation matrix about the axis along the vector $\boldsymbol{L}$ can be written in the form

$$R(\boldsymbol{L}, \tfrac{1}{2}\varepsilon) = \sigma_0 - (\tfrac{1}{2}\varepsilon)(L_1\sigma_1 + L_2\sigma_2 + L_3\sigma_3). \tag{1.3.13}$$

Let us define a vector $\boldsymbol{\sigma}$ having as its components the Pauli matrices

$$\boldsymbol{\sigma} = (\sigma_1, \sigma_2, \sigma_3). \tag{1.3.14}$$

The expression $(L_1\sigma_1 + L_2\sigma_2 + L_3\sigma_3)$ can then be written in the form of a 'scalar product' which represent a sum of matrices

$$L_1\sigma_1 + L_2\sigma_2 + L_3\sigma_3 = \boldsymbol{L}\cdot\boldsymbol{\sigma}. \tag{1.3.15}$$

The restricted expansion (1.3.4) of the rotation matrix may then be written

$$R(\boldsymbol{L}, \tfrac{1}{2}\varepsilon) = \sigma_0 - (i\tfrac{1}{2}\varepsilon)\boldsymbol{L}\cdot\boldsymbol{\sigma}. \tag{1.3.16}$$

The rotation matrix (1.3.1) for a finite angle $\frac{1}{2}\theta$ may then be written in the remarkable form

$$R(\boldsymbol{L}, \tfrac{1}{2}\theta) = \cos\tfrac{1}{2}\theta\sigma_0 - i\sin\tfrac{1}{2}\theta\boldsymbol{L}\cdot\boldsymbol{\sigma}. \tag{1.3.17}$$

Thus we see how the Pauli matrices are introduced in a fundamental way in the theory of spinors.

1.3.3 *Properties of the Pauli Matrices*

Anticommutativity

The squares of the Pauli matrices are equal to the unit matrix σ_0, that is

$$\sigma_1^2 = \sigma_2 = \sigma_3^2 = \sigma_0. \tag{1.3.18}$$

These matrices satisfy the following ANTICOMMUTATION RELATIONS

$$\sigma_i\sigma_j = -\sigma_j\sigma_i \qquad \text{for} \quad i \neq j, \quad i, j = 1, 2, 3. \tag{1.3.19}$$

We say that the Pauli matrices *anticommute*. Relations (1.3.18) and (1.3.19) can be brought together into the following form

$$\sigma_i\sigma_j + \sigma_j\sigma_i = 2\delta_{ij}\sigma_0, \qquad i, j = 1, 2, 3. \tag{1.3.20}$$

The products of the matrices with each other give the following relations

$$\sigma_1\sigma_2 = i\sigma_3, \qquad \sigma_2\sigma_3 = i\sigma_1, \qquad \sigma_3\sigma_1 = i\sigma_2. \tag{1.3.21}$$

Relations (1.3.19) and (1.3.21) allow us to obtain the following COMMUTATION RELATIONS

$$\begin{aligned} \sigma_1\sigma_2 - \sigma_2\sigma_1 &= 2i\sigma_3, \\ \sigma_2\sigma_3 - \sigma_3\sigma_2 &= 2i\sigma_1, \\ \sigma_3\sigma_1 - \sigma_1\sigma_3 &= 2i\sigma_2. \end{aligned} \tag{1.3.22}$$

The Vector Space of 2×2 *Matrices*

The Pauli matrices and the unit matrix σ_0 form a set of *linearly independent* square matrices. In fact, let us write that the sum of the following matrices is equal to the null matrix

$$\lambda_0\sigma_0 + \lambda_1\sigma_1 + \lambda_2\sigma_2 + \lambda_3\sigma_3 = \mathbf{0}. \tag{1.3.23}$$

This matrix equation is equivalent to the following equations obtained by writing out the matrices σ_i

$$\begin{aligned} \lambda_0 + \lambda_3 = 0, &\qquad \lambda_1 - i\lambda_2 = 0, \\ \lambda_1 + i\lambda_2 = 0, &\qquad \lambda_0 - \lambda_3 = 0. \end{aligned} \tag{1.3.24}$$

These equations can only be satisfied by the solution

$$\lambda_0 = \lambda_1 = \lambda_2 = \lambda_3 = 0. \tag{1.3.25}$$

The four matrices σ_i are thus linearly independent and are able to generate the 2×2 matrices with real or complex elements. The matrices σ_i thus form a basis of the vector space of matrices of second order over the real or complex numbers. It is easily verified that every matrix

$$A = \begin{bmatrix} a & b \\ c & d \end{bmatrix} \tag{1.3.26}$$

can be written in the form

$$A = \alpha\sigma_0 + \beta\sigma_1 + \gamma\sigma_2 + \delta\sigma_3 \tag{1.3.27}$$

with

$$\alpha = \tfrac{1}{2}(a+d), \qquad \beta = \tfrac{1}{2}(b+c), \qquad \gamma = \tfrac{1}{2}i(b-c), \qquad \delta = \tfrac{1}{2}(a-d).$$

Isomorphic Spaces

Let $\boldsymbol{X}$ be an arbitrary vector of the vector space E_3 with $\{\boldsymbol{e}_1, \boldsymbol{e}_2, \boldsymbol{e}_3\}$ as its basis. We have

$$\boldsymbol{X} = x\boldsymbol{e}_1 + y\boldsymbol{e}_2 + z\boldsymbol{e}_3. \tag{1.3.28}$$

To every vector $\boldsymbol{X}$ we can make correspond a 2×2 matrix such that

$$X = x\sigma_1 + y\sigma_2 + z\sigma_3. \tag{1.3.29}$$

This correspondence is an isomorphism between the vector space E_3 and a part of the vector space of matrices of order 2. These matrices have the expression

$$X = \begin{bmatrix} z & x - iy \\ x + iy & -z \end{bmatrix}. \tag{1.3.30}$$

This latter expression for X is found in the relation linking an isotropic vector with components (x, y, z) and its associated spinor (ψ, ϕ). In fact, from relations (1.1.22) and (1.1.24) we obtain the new relations

$$\begin{aligned} z\psi + (x - iy)\phi &= 0, \\ (x + iy)\psi - z\phi &= 0. \end{aligned} \tag{1.3.31}$$

When put into matrix form these latter two relations may be written

$$\begin{bmatrix} z & x - iy \\ x + iy & -z \end{bmatrix} \begin{bmatrix} \psi \\ \phi \end{bmatrix} = \begin{bmatrix} 0 \\ 0 \end{bmatrix}. \tag{1.3.32}$$

1.4 Algebraic Properties of Spinors

1.4.1 Operations on Spinors

Addition and Multiplication by a Scalar

Pairs of complex numbers ψ and ϕ are going to be able to be considered as vectors if we define on these pairs the classical operations of addition of the pairs and multiplication by a scalar.

By definition, for two pairs $\boldsymbol{x} = (\psi_1, \phi_1)$ and $\boldsymbol{y} = (\psi_w, \phi_2)$ addition brings into correspondence another pair $\boldsymbol{z} = (\psi_3, \phi_3)$, denoted $\boldsymbol{x} + \boldsymbol{y}$, such that

$$\begin{aligned} \boldsymbol{x} + \boldsymbol{y} &= (\psi_1 + \psi_2, \phi_1 + \phi_2) \\ &= (\psi_3, \phi_3) = \boldsymbol{z}. \end{aligned} \tag{1.4.1}$$

The pair (ψ_3, ϕ_3) is called the sum of $\boldsymbol{x} + \boldsymbol{y}$.

Also, by definition, for a pair $\boldsymbol{x} = (\psi, \phi)$ multiplication by a complex number λ brings into correspondence another pair $\boldsymbol{w} = (\psi', \phi')$, denoted $\lambda\boldsymbol{x}$, such that

$$\lambda\boldsymbol{x} = (\lambda\psi, \lambda\phi) = (\psi', \phi') = \boldsymbol{w}. \tag{1.4.2}$$

The pair (ψ', ϕ') is called the product of $\boldsymbol{x}$ by λ. These two operations on pairs of complex numbers bring another element of this set of pairs of complex numbers into correspondence with one or more elements of the set.

Properties of Operations on Spinors

Let us consider more specifically the pairs of complex numbers forming spinors and on which the two preceding operations are defined. We can verify that the set of spinors is stable under these two operations, that is to say that the sum of two spinors gives another spinor, and that the product of a spinor with a complex number also gives a spinor. If we take as the definition of spinors the one which is based on their rotation property in the spinor space and characterized by the matrix $R(\boldsymbol{L}, \frac{1}{2}\theta)$ given by (1.2.13), this verification is immediate.

In fact, under an arbitrary rotation of the coordinate system of three-dimensional space a spinor (ψ_1, ϕ_1) undergoes a unitary transformation which, using expressions (1.2.13) for the rotation matrix, can be written

$$\psi_1' = a\psi_1 - b^*\phi_1, \qquad \phi_1' = \psi_1 + a^*\phi_1. \tag{1.4.3}$$

Let us consider a second spinor (ψ_2, ϕ_2) which is transformed similarly. As a result the transform of the sum of these two spinors may be written

$$\begin{aligned}(\psi_1' + \psi_2') &= a(\psi_1 + \psi_2) - b^*(\phi_1 + \phi_2), \\ (\phi_1' + \phi_2') &= b(\psi_1 + \psi_2) + a^*(\phi_1 + \phi_2).\end{aligned} \tag{1.4.4}$$

Thus the sum of two spinors transforms as a spinor, and thus this sum is also a spinor. The same is true for multiplication of a spinor by a complex number.

1.4.2 Properties of Operations on Spinors

The addition of spinors and their multiplication by a scalar satisfy the following properties

Addition of Spinors

- Commutativity: $\boldsymbol{x} + \boldsymbol{y} = \boldsymbol{y} + \boldsymbol{x}$.
- Associativity: $\boldsymbol{x} + (\boldsymbol{y} + \boldsymbol{z}) = (\boldsymbol{x} + \boldsymbol{y}) + \boldsymbol{x}$.
- There exists a null spinor, denoted $\mathbf{0} = (0, 0)$, such that $\boldsymbol{x} + \mathbf{0} = \boldsymbol{x}$.
- For an arbitrary spinor $\boldsymbol{x}$ there exists a spinor denoted $(-\boldsymbol{x})$, called its opposite, such that $\boldsymbol{x} + (-\boldsymbol{x}) = \mathbf{0}$.

Multiplication by a Complex Number

- Associativity: $\lambda_1(\lambda_2\boldsymbol{x}) = (\lambda_1\lambda_2)\boldsymbol{x}$.
- Distributivity with respect to scalar addition

$$(\lambda_1 + \lambda_2)\boldsymbol{x} = \lambda_1\boldsymbol{x} + \lambda_2\boldsymbol{x}.$$

— Distributivity with respect to spinor addition

$$\lambda(\boldsymbol{x}+\boldsymbol{y}) = \lambda\boldsymbol{x} + \lambda\boldsymbol{y}.$$

— For the scalar 1 there holds $1\boldsymbol{x} = \boldsymbol{x}$.

These various properties are proved easily from definitions (1.4.1) and (1.4.2) of the operations. The set of spinors provided with these two laws of composition thus form a vector space over the complex numbers. The spinors are thus special vectors.

1.4.3 The Basis of the Vector Space of Spinors

Linearly Independent Spinors

Two spinors (ψ_1, ϕ_1) and (ψ_2, ϕ_2) are linearly independent if

$$\lambda_1(\psi_1, \phi_1) + \lambda_2(\psi_2, \phi_2) = (0, 0), \tag{1.4.5}$$

that is to say, if there simultaneously hold

$$\lambda_1\psi_1 + \lambda_2\psi_2 = 0, \qquad \lambda_1\phi_1 + \lambda_2\phi_2 = 0. \tag{1.4.6}$$

This system of equations has vanishing solutions λ_1 and λ_2 if the following determinant is different from zero

$$\begin{aligned}\begin{vmatrix} \psi_1 & \psi_2 \\ \phi_1 & \phi_2 \end{vmatrix} &= \psi_1\phi_2 - \phi_1\phi_2 \\ &\neq 0.\end{aligned} \tag{1.4.7}$$

In this case the spinors are linearly independent. In order that the determinant be different from zero it suffices to consider a first spinor $(\psi_1, \phi_1) = (\psi, \phi)$, the components of which are not both identically zero, and thus such that:

$$\psi\psi^* + \phi\phi^* \neq 0, \tag{1.4.8}$$

and to take a second spinor (ψ_2, ϕ_2) such that

$$(\psi_2, \phi_2) = (-i\phi^*, i\psi^*). \tag{1.4.9}$$

The determinant (1.4.6) of the system may then be written for these spinors

$$\begin{aligned}\begin{vmatrix} \psi & i\phi^* \\ \phi & i\psi^* \end{vmatrix} &= i(\psi\psi^* + \phi\phi^*) \\ &\neq 0,\end{aligned} \tag{1.4.10}$$

and it is different from zero as a result of the hypothesis (1.4.8). Consequently the two spinors

$$(\psi, \phi) \quad \text{and} \quad (-i\phi^*, i\psi^*) \tag{1.4.11}$$

are linearly independent. The spinors (ψ, ϕ) and $(-i\phi^*, i\psi^*)$ are able to generate the spinor space and they form a basis of the spinor vector space. Every spinor (η, ξ) can be written in the form of a linear combination of basis spinors

$$(\eta, \xi) = \alpha(\psi, \phi) + \beta(-i\phi^*, i\psi^*). \tag{1.4.12}$$

The complex numbers α and β are the components of the spinor.

The Natural Basis

We can consider for the basis of the spinor vector space the following pairs

$$\boldsymbol{e}_1 = (1, 0), \qquad \boldsymbol{e}_2 = (0, 1). \tag{1.4.13}$$

These two vectors appear to be identical to those which are able to serve as the basis vector of a plane, for example. However, considered as basis vectors of the spinor space the vectors $\boldsymbol{e}_1$ and $\boldsymbol{e}_2$ transform under a rotation in spinor space, by hypothesis, according to the matrix $R(\boldsymbol{L}, \frac{1}{2}\alpha)$, and this transformation leads to particular complex numbers.

Every spinor $\boldsymbol{z} = (\eta, \zeta)$ can be written in the basis $\{\boldsymbol{e}_1, \boldsymbol{e}_2\}$, that is to say,

$$\boldsymbol{z} = (\eta, \zeta) = \eta(1, 0) + \zeta(0, 1). \tag{1.4.14}$$

In such a basis the components of the spinor (η, ζ) are identical to the complex numbers which define the spinor. The basis $\{\boldsymbol{e}_1, \boldsymbol{e}_2\}$ is the NATURAL BASIS.

1.4.4 Hermitian Vector Spaces

The vector space of spinors, denoted $\mathcal{S}$, can be given a Hermitian product. Let us recall that the latter is a mapping which associates with every pair $(\boldsymbol{x}, \boldsymbol{y})$ of elements of the vector space $\mathcal{S}$ a complex number, denoted $\boldsymbol{x}\cdot\boldsymbol{y}$, satisfying the following properties:

— For two arbitrary spinors $\boldsymbol{x}, \boldsymbol{y}$ of the space $\mathcal{S}$ we have

$$\boldsymbol{x}\cdot\boldsymbol{y} = (\boldsymbol{y}\cdot\boldsymbol{x})^*, \tag{1.4.15}$$

where $(\boldsymbol{y}\cdot\boldsymbol{x})^*$ denotes the complex conjugate of the number $(\boldsymbol{x}\cdot\boldsymbol{x})$.

— If $\boldsymbol{x}, \boldsymbol{y}, \boldsymbol{z}$ are spinors of $\mathcal{S}$ then

$$\boldsymbol{x} \cdot (\boldsymbol{y} + \boldsymbol{z}) = \boldsymbol{z} \cdot \boldsymbol{y} + \boldsymbol{x} \cdot \boldsymbol{z}. \tag{1.4.16}$$

— If λ is a complex number, we have

$$(\lambda \boldsymbol{x}) \cdot \boldsymbol{y} = \lambda (\boldsymbol{x} \cdot \boldsymbol{y}). \tag{1.4.17}$$

From the properties (1.4.15) and (1.4.17) we deduce that

$$\boldsymbol{x} \cdot (\lambda \boldsymbol{y}) = \lambda^* (\boldsymbol{x} \cdot \boldsymbol{y}). \tag{1.4.18}$$

— The Hermitian product is *definite* if

$$\boldsymbol{x} \cdot \boldsymbol{x} \neq 0 \qquad \text{for all} \quad \boldsymbol{x} \neq 0. \tag{1.4.19}$$

The Hermitian product is POSITIVE DEFINITE if $\boldsymbol{x} \cdot \boldsymbol{x} \geqslant 0$ for every spinor $\boldsymbol{x}$ and if $\boldsymbol{x} \cdot \boldsymbol{x} > 0$ when $\boldsymbol{x} \neq 0$. The Hermitian product is *negative definite* if $\boldsymbol{x} \cdot \boldsymbol{x} < 0$. A vector space with a positive definite Hermitian product is called a HERMITIAN VECTOR SPACE. The name Hermitian product is often kept without any other indication if the product is positive definite, and this is what we shall do in the remainder.

The Classical Hermitian Product

For spinors we shall use the Hermitian product of two spinors $\boldsymbol{x} = (\psi, \phi)$ and $\boldsymbol{y} = (\eta, \xi)$ defined thus

$$\boldsymbol{x} \cdot \boldsymbol{y} = \psi \eta^* + \phi \zeta^*. \tag{1.4.20}$$

It is easily verified that such a product satisfies the Hermitian product's axioms (1.4.15)–(1.4.19) This product is certainly positive definite, for if $\boldsymbol{x}$ is different from the null spinor then at least ψ or ϕ are different from zero, and as a result

$$\boldsymbol{x} \cdot \boldsymbol{x} = \psi \psi^* + \phi \phi^* > 0.$$

1.4.5 Properties of the Hermitian Product

The classical definitions of the properties of the scalar product keep the same meanings for the Hermitian product. It will be said, for example, that two spinors $\boldsymbol{x}$ and $\boldsymbol{y}$ are mutually *orthogonal* if

$$\boldsymbol{x} \cdot \boldsymbol{y} = 0. \tag{1.4.21}$$

Norm of a Spinor

The Hermitian product of a spinor $\boldsymbol{x} = (\psi, \phi)$ with itself allows us to define its NORM, denoted $\|\boldsymbol{x}\|$, which is the number

$$\|\boldsymbol{x}\| = \sqrt{\boldsymbol{x}\cdot\boldsymbol{x}} = \sqrt{\psi\psi^* + \phi\phi^*}. \tag{1.4.22}$$

Since $\boldsymbol{x}\cdot\boldsymbol{x}$ is positive the square root of $\boldsymbol{x}\cdot\boldsymbol{x}$ is a real number. For every spinor $\lambda\boldsymbol{x}$, where λ is a complex number, we have

$$\begin{aligned} \|\lambda\boldsymbol{x}\|^2 &= (\lambda\boldsymbol{x})\cdot(\lambda\boldsymbol{x}) \\ &= \lambda\lambda^*(\boldsymbol{x}\cdot\boldsymbol{x}) \\ &= |\lambda|^2(\boldsymbol{x}\cdot\boldsymbol{x}). \end{aligned} \tag{1.4.23}$$

The square root of the last term of the previous equation gives us

$$\|\lambda\boldsymbol{x}\| = |\lambda|\,\|\boldsymbol{x}\|. \tag{1.4.24}$$

Unitary Spinors

It will be said that an arbitrary spinor $\boldsymbol{e}$ is a unitary spinor if

$$\|\boldsymbol{e}\| = 1. \tag{1.4.25}$$

Given an arbitrary nonzero spinor $\boldsymbol{x}$ we obtain a unitary spinor by dividing it by its norm, that is,

$$\boldsymbol{e} = \frac{\boldsymbol{x}}{\|\boldsymbol{x}\|}. \tag{1.4.26}$$

The spinor $\boldsymbol{e}$ then has norm unity,

$$\|\boldsymbol{e}\| = 1.$$

Orthonormed Basis

The spinors $\boldsymbol{e}_1 = (1, 0)$ and $\boldsymbol{e}_2 = (0, 1)$ are clearly unitary and an orthonormed basis of the vector space of spinors.

Let us consider another basis formed from a unitary spinor $\boldsymbol{e}_1 = (\psi, \phi)$. We have

$$\begin{aligned} \|\boldsymbol{e}_1\| &= \sqrt{\boldsymbol{e}_1\cdot\boldsymbol{e}_1} \\ &= \sqrt{\psi\psi^* + \phi\phi^*} \\ &= 1. \end{aligned} \tag{1.4.27}$$

We have seen beforehand that the spinor $\boldsymbol{e}_2 = (-i\phi^*, i\psi^*)$ is linearly independent of the spinor $\boldsymbol{e}_1$. As this latter is unitary the spinor $\boldsymbol{e}_2$ is unitary also. In fact, we have

$$\begin{aligned} \boldsymbol{e}_2 \cdot \boldsymbol{e}_2 &= (-i\phi^*)(i\phi) + (i\psi^*)(-i\psi) \\ &= \phi\phi^* + \psi^*\psi \\ &= 1. \end{aligned} \tag{1.4.28}$$

Let us calculate the Hermitian product $\boldsymbol{e}_1 \cdot \boldsymbol{e}_2$. We obtain

$$\begin{aligned} \boldsymbol{e}_1 \cdot \boldsymbol{e}_2 &= (\psi, \phi) \cdot (-i\phi^*, i\psi^*) \\ &= \psi(i\phi) + \phi(-i\psi) \\ &= 0. \end{aligned} \tag{1.4.29}$$

The spinors $\boldsymbol{e}_1$ and $\boldsymbol{e}_2$ are mutually orthogonal. We have seen that they form a basis of the vector space of spinors. Such a basis is therefore orthonormal.

1.4.6 The Use of an Antisymmetric Metric Tensor

Instead of providing the vector space of spinors with the classical metric defined by the Hermitian product (1.4.20) we can a priori introduce other metrics. Thus, since a spinor is associated with an isotropic vector $\boldsymbol{Z}$ of zero norm and defined by (1.1.18), it has led to the introduction of a particular metric which also gives spinors a zero norm.

To do so we define the metric of the spinor space by the following antisymmetric metric tensor

$$\begin{bmatrix} g_{11} & g_{12} \\ g_{21} & g_{22} \end{bmatrix} = \begin{bmatrix} 0 & 1 \\ 1 & 0 \end{bmatrix}. \tag{1.4.30}$$

Let $\boldsymbol{e}_1$ and $\boldsymbol{e}_2$ be the spinors of a basis such that $g_{ij} = \boldsymbol{e}_1 \cdot \boldsymbol{e}_2$. Let us consider two spinors decomposed in this basis

$$\boldsymbol{x} = \psi \boldsymbol{e}_1 + \phi \boldsymbol{e}_2, \qquad \boldsymbol{y} = \eta \boldsymbol{e}_1 + \zeta \boldsymbol{e}_2. \tag{1.4.31}$$

The 'scalar product' of these spinors may then be written, on using the classical expression and taking account of (1.4.31)

$$\begin{aligned} \boldsymbol{x} \cdot \boldsymbol{y} &= \psi\eta g_{11} + \psi\zeta g_{12} + \phi\eta g_{21} + \phi\zeta g_{22} \\ &= \psi\zeta - \phi\eta. \end{aligned} \tag{1.4.32}$$

Consequently, the 'scalar product' of a spinor with itself is zero

$$\boldsymbol{x} \cdot \boldsymbol{x} = \psi\phi - \phi\psi = 0. \tag{1.4.33}$$

The 'norm' of the spinor, equal to $\sqrt{\boldsymbol{x} \cdot \boldsymbol{x}}$, is thus zero for the metric imposed on the space of spinors, the latter thus becoming isotropic vectors.

1.5 Solved Problems

EXERCISE **1.1**

Consider a unitary spinor (ψ, ϕ) defined by (1.1.6) and (1.1.9) where the coordinates x, y, z are those of a point M of three-dimen- sional space.

1. Show that the point $M(x, y, z)$ lies on a sphere of unit radius.

2. Consider the following transformation U

$$\psi' = a\psi + b\phi, \qquad \phi' = -b^*\psi + a^*\phi,$$

   where a and b are complex parameters satisfying $aa^* + bb^* = 1$. Show that the matrix $M(U)$ of this transformation is unitary.

3. Show that the transformation U takes a point $M(x, y, z)$ of the unit sphere into another point $M'(x', y', z')$ of the same sphere.

4. Show that the transformation which takes the point $M(x, y, z)$ into the point $M'(x', y', z')$ on the unit sphere is a rotation R in three-dimensional space.

5. Show that two transformations U and $-U$ correspond to every rotation R in three-dimensional space.

SOLUTION **1.1**

1. As the spinor is unitary we have $\psi\psi^* + \phi\phi^* = 1$. Using (1.1.6) and (1.1.9) gives us

$$x^2 + y^2 + z^2 = (\psi\psi^* + \phi\phi^*)^2 = 1.$$

   The point $M(x, y, z)$ lies on the sphere of unit radius.

2. A matrix $M(U)$ is said to be unitary if

$$M(U)M(U)^\dagger = M(U)^\dagger M(U) = 1\!\!1,$$

   where $M(U)^\dagger$ is the adjoint matrix of $M(U)$ and $1\!\!1$ is the unit matrix.

   As the adjoint matrix is the transpose of the conjugate, we have

$$M(U) = \begin{bmatrix} a & b \\ -b^* & a^* \end{bmatrix},$$

   whence

$$M(U)^\dagger \begin{bmatrix} a^* & -b \\ b^* & a \end{bmatrix}.$$

The defining equations of a unitary matrix are then easily verified taking the equation $aa^* + bb^* = 1$ into account

$$\begin{bmatrix} a & b \\ -b^* & a^* \end{bmatrix} \begin{bmatrix} a^* & -b \\ b^* & a \end{bmatrix} = \begin{bmatrix} a^* & -b \\ b^* & a \end{bmatrix} \begin{bmatrix} a & b \\ -b^* & a^* \end{bmatrix} = \begin{bmatrix} 1 & 0 \\ 0 & 1 \end{bmatrix}.$$

3. The norm of the transformed spinor (ψ', ϕ') is preserved, since on taking $\psi\psi^* + \phi\phi^* = 1$ and $aa^* + bb^* = 1$ into account

$$\begin{aligned} \psi'\psi'^* + \phi'\phi'^* &= x'^2 + y'^2 + z'^2 \\ &= 1. \end{aligned}$$

The point $M(x, y, z)$ is thus transformed into another point $M'(x', y', z')$ of the unit sphere.

4. The two points M, M' and the center O of the sphere define a plane. The intersection of this plane with the sphere is a circle along which the rotation R is carried out which allows us to go from the point M to the point M'.

An analytical proof is also possible, but it is more fastidious. The transform $M'(x', y', z')$ of the point $M(x, y, z)$ has, by (1.1.6) and (1.1.9), as its coordinates

$$\begin{aligned} x' &= \psi'\phi'^* + \psi'^*\phi', \\ y' &= i(\psi'\phi'^* - \psi'^*\phi'), \\ z' &= \psi'\psi'^* - \phi'\phi'^*. \end{aligned}$$

Replacing ψ' and ϕ' in these expressions by

$$\begin{aligned} \psi' &= a\psi + b\phi, \\ \phi' &= -b^*\psi + a^*\phi, \end{aligned}$$

and expanding, we obtain the values of x', y', z' as functions of ψ and ϕ, then as functions of x, y, z. The matrix of the transformation obtained in this way is an orthogonal matrix with determinant equal to unity corresponding to a rotation in three-dimensional space. By analogy it will be said that the transformation U is a 'rotation' in the space of spinors.

5. Relations (1.1.6) and (1.1.9) between ψ, ϕ and x, y, z show that the coordinates are sums or differences of products of the components ψ, ϕ and their conjugates. If a and b are changed into $-a$ and $-b$ in the transformation U, which thereupon becomes $-U$, the transformed coordinates x', y', z' remain unchanged. As a result, to every rotation on the unit sphere there correspond the two transformations U and $-U$.

EXERCISE **1.2**

The lines with coefficient equal to $\pm i$ of a plane referred to two rectangular axes are called isotropic lines.

1. Write down the equations of the isotropic lines in a plane xOy.
2. Show that the angular coefficient of an isotropic line is invariant under every change of axes of rectangular coordinates.
3. Let $M_1 = (x_1, y_1)$ and $M_2 = (x_2, y_2)$ be two points of the same isotropic line. Show that the distance between these points is zero.
4. Let a vector $\boldsymbol{X} = (x, y)$ lie along an isotropic line. Determine the length of $\boldsymbol{X}$.

SOLUTION **1.2**

1. A line is a set of real or imaginary points of which the coefficients satisfy a first-order equation

$$ax + by + c = 0$$

which has real or imaginary coefficients.

The isotropic lines can be represented by the equations

$$y \pm ix + \lambda = 0.$$

2. As a translation of the axes leaves the angular coefficient of an arbitrary line invariant, it is necessary to show that the coefficient of an isotropic line is invariant under a rotation of the rectangular axes about the origin. To do so it suffices to consider an isotropic line issuing from the origin

$$y = \pm ix.$$

The transformation formulas of the coordinates for a rotation α have the form

$$\begin{aligned} x &= x' \cos\alpha - y' \sin\alpha, \\ y &= x' \sin\alpha + y' \cos\alpha. \end{aligned}$$

The equation of the isotropic line in the new system $x'Oy'$ is

$$x' \sin\alpha + y' \cos\alpha = \pm i(x' \cos\alpha - y' \sin\alpha),$$

so

$$y'(\cos\alpha \pm i \sin\alpha) = \pm ix'(\cos\alpha \pm i \sin\alpha),$$

whence

$$y' = \pm ix'.$$

In the new system the isotropic line again has $\pm i$ as its angular coefficient.

3. If the two points M_1 and M_2 lie on the same isotropic line we have

$$y_1 = \pm ix_1, \qquad y_2 = \pm ix_2,$$

and so

$$y_2 - y_1 = \pm i(x_2 - x_1).$$

As the angular coefficient is equal to $+i$ or $-i$ we have the two cases

$$(y_2 - y_1)^2 = -(x_2 - x_1)^2,$$

that is, a distance

$$(y_2 - y_1)^2 + (x_2 - x_1)^2 = 0.$$

4. Let M_1 and M_2 be the extremities of a vector $\boldsymbol{X}$ such that $\boldsymbol{X} = \boldsymbol{M}_1\boldsymbol{M}_2$. Then

$$y = y_2 - y_1, \qquad x = \pm i(x_2 - x_1).$$

The Hermitian product of $\boldsymbol{X}$ with itself gives the square of its norm, that is

$$\begin{aligned}
\langle \boldsymbol{X}, \boldsymbol{X} \rangle &= \|\boldsymbol{X}\|^2 \\
&= yy^* + xx^* \\
&= (y_2 - y_1)^2 + (x_2 - x_1)^2 \\
&= 0.
\end{aligned}$$

A vector along an isotropic line has zero length; it is an isotropic vector.

EXERCISE **1.3**
According to relations (1.1.6) and (1.1.9) the vector $\boldsymbol{OP}$ of Figure 1.1 has as components

$$\begin{aligned} x &= \psi\phi^* + \psi^*\phi, \\ y &= i(\psi\phi^* - \psi^*\phi), \\ z &= \psi\psi^* - \phi\phi^*. \end{aligned}$$

Using the components of the spinors (ψ, ϕ) and (ψ^*, ϕ^*) put into matrix form, write down the components of the vector $\boldsymbol{OP}$ in the form of products of matrices using the Pauli matrices.

SOLUTION **1.3**
The component x of the vector $\boldsymbol{OP}$ can be put into the following matrix form

$$\begin{aligned} x &= \psi\phi^* + \psi^*\phi \\ &= \begin{bmatrix} \psi^* & \phi^* \end{bmatrix} \begin{bmatrix} \psi \\ \phi \end{bmatrix}, \\ y &= i(\psi\phi^* - \psi^*\phi) \\ &= \begin{bmatrix} \psi^* & \phi^* \end{bmatrix} \begin{bmatrix} \psi \\ \phi \end{bmatrix}, \\ z &= \psi\psi^* - \phi\phi^* \\ &= \begin{bmatrix} \psi^* & \phi^* \end{bmatrix} \begin{bmatrix} \psi \\ \phi \end{bmatrix}. \end{aligned}$$

Let us write

$$\chi = \begin{bmatrix} \psi \\ \phi \end{bmatrix} \qquad \text{and} \qquad \chi^\dagger = \begin{bmatrix} \psi^* & \phi^* \end{bmatrix}.$$

On the other hand, let $\sigma_1, \sigma_2, \sigma_3$ be the Pauli matrices given by (1.3.12). The matrix products may then be written

$$x = \chi^\dagger \sigma_1 \zeta, \qquad y = \chi^\dagger \sigma_2 \chi, \qquad z = \chi^\dagger \sigma_3 \chi.$$

The Pauli matrices were introduced into the theory of spinors in a fundamental way. Each matrix corresponds to a component along one of the axes of the reference trihedron. We have seen that they are the infinitesimal rotation matrices about the axes Ox, Oy, Oz to within a multiplicative factor.

2

Spinors and $SU(2)$ Group Representations

In the course of this chapter we assume that the reader already knows the principal notions of the theory of finite groups and their representations. If this is not the case we refer the reader to the Appendix, where we have brought together the definitions it is necessary to know as well as different examples of finite groups which also allow a familiarization with those notions.

In this chapter we give, first of all, some items which will be useful to us for continuous groups. These notions will allow us to replace the results obtained in the previous chapter in the framework of the theory of groups. This generalization will allow us to obtain spinors of arbitrary order attached to the unimodular unitary group.

2.1 Lie Groups

2.1.1 Examples of Continuous Groups

The examples given in the Appendix essentially concern groups with a finite number of elements when the continuous groups are groups, characterized by one or more continuous parameters, which have an infinite number of elements. This is the case of groups of rotations in the plane, denoted $\boldsymbol{SO}(2)$, and in three-dimensional space, denoted $\boldsymbol{SO}(3)$. These groups have a continuous infinity of rotations.

REMARK **2.1.** The term 'continuous groups' is not an expression used by mathematicians, who call them *Lie groups*, and of which we shall speak later. However, as the name 'continuous groups' is often used in physics and is evocative, we shall keep its use here. ■

The Group of Rotations in a Plane

Let us consider by way of an example the group $\boldsymbol{SO}(2)$ of rotations $R(\alpha)$ through an arbitrary angle α about a given axis. This group is that of rotations in a plane. The angle α is the continuous parameter which characterizes the elements of the group. The domain of variation of α is bounded, $0 \leqslant \alpha < 2\pi$, and it is said that the group is a COMPACT GROUP.

These rotations can be considered as mappings transforming a point $M(x, y)$ of a two-dimensional space into another point $M'(x', y')$ of the same space, such that

$$\begin{aligned} x' &= x \cos\alpha - y \sin\alpha \\ &= f_1(x, y; \alpha), \\ y' &= x \sin\alpha + y \cos\alpha \\ &= f_2(x, y; \alpha). \end{aligned} \tag{2.1.1}$$

Two functions $f_1(x, y; \alpha)$ and $f_2(x, y; \alpha)$ are obtained which define analytically the rotation $R(\alpha)$. The product of two successive rotations $R(\alpha_1)$ and $R(\alpha_2)$ is a rotation belonging to the group and such that

$$\begin{aligned} R(\alpha_1)R(\alpha_2) &= R(\alpha_1 + \alpha_2) \\ &= R(\alpha_3). \end{aligned} \tag{2.1.2}$$

The group's law is characterized by a function ϕ such that

$$\phi(\alpha_1, \alpha_2) = \alpha_1 + \alpha_2 = \alpha_3. \tag{2.1.3}$$

The identity transformation E corresponds to $\alpha_0 = 2k\pi$, and we can choose $\alpha_0 = 0$. For the multiplication law of the group we have

$$\phi(\alpha, 0) = \alpha.$$

The inverse transformation $R^{-1}(\alpha)$ is equal to $R(-\alpha)$ and gives

$$\phi(\alpha, -\alpha) = 0.$$

The Group of Spacial Rotations

The group of spacial rotations $\boldsymbol{SO}(3)$ consists of all the rotations about any axis passing through a given point. Each element of the group can be characterized by three continuous parameters. In fact, an arbitrary rotation can be decomposed into three successive rotations through angles α, β, γ, respectively about the axes Ox, Oy, Oz of a Cartesian frame of reference. The domains of variation of the parameters α, β, γ are bounded, and the group is said to be compact.

These rotations transform a point $M(x, y, z)$ into another point $M'(x', y', z')$ of a three-dimensional space, and we have three functions

$$\begin{aligned} x' &= f_1(x, y, z; \alpha, \beta, \gamma), \\ y' &= f_2(x, y, z; \alpha, \beta, \gamma), \\ z' &= f_3(x, y, z; \alpha, \beta, \gamma), \end{aligned} \tag{2.1.4}$$

which define the rotation $R(\alpha, \beta, \gamma)$ analytically.

2.1.2 Analytic Definition of Continuous Groups

The elements of a continuous group are, quite generally, mappings which transform a point $M(x_1, x_2, \ldots, x_n)$ of an n-dimensional space into another point $M'(x_1', x_2', \ldots, x_n')$ of the same space. These transformations depend upon r continuous parameters $\alpha_1, \alpha_2, \ldots, \alpha_r$. Thus we have n functions

$$x_i' = f_i(x_1, \ldots, x_n; \alpha_1, \ldots, \alpha_r) \tag{2.1.5}$$

which define these transformations analytically and which are denoted $G(\alpha_1, \ldots, \alpha_r)$. In the remainder we shall consider transformations (2.1.5) which are *linear*, and we shall restrict ourselves to linear groups.

The set of transformations corresponding to all possible values of the parameters α_j in a given continuous domain forms a group if it has the following properties:

— The *product* of two successive transformations (the name given to the transformation which is the composition of the two) is a transformation of this set, and such that

$$G(\alpha_1, \ldots, \alpha_r) G(\alpha_1', \ldots, \alpha_r') = G(\alpha_1'', \ldots, \alpha_r''). \tag{2.1.6}$$

We shall write as ϕ_i the functions defined by

$$\alpha_i'' = \phi_i(\alpha_1, \ldots, \alpha_r; \alpha_1', \ldots, \alpha_r'), \tag{2.1.7}$$

with $i = 1, \ldots, r$. These functions will be assumed to be differentiable with respect to all the variables.

— The product of two transformations must be associative.

— The identity transformation E, defined by $x_i{}' = x_i$, must be able to be obtained. It satisfies

$$\begin{aligned} G(\alpha_1, \ldots, \alpha_r)E &= EG(\alpha_1, \ldots, \alpha_r) \\ &= G(\alpha_1, \ldots, \alpha_r), \end{aligned}$$

and we write as α_j^0 the corresponding values of the parameters, such that

$$x_i = f_i(x_i, \ldots, x_n; \alpha_1^0, \ldots, \alpha_r^0).$$

As the parameters can be modified by a change of origin we can set $\alpha_j{}^0 = 0$, whence

$$x_i = f_i(x_1, \ldots, x_n; 0, \ldots, 0). \tag{2.1.8}$$

For the group's multiplication law we have

$$\phi(\alpha_1, \ldots, \alpha_r; 0, \ldots, 0) = (\alpha_1, \ldots, \alpha_r). \tag{2.1.9}$$

— Each transformation $G(\alpha_1, \ldots, \alpha_r)$ must be invertible and its inverse must belong to the set of transformations. We write the inverse transformation $G^{-1}(\alpha_1, \ldots, \alpha_r)$ and we have

$$G(\alpha_1, \ldots, \alpha_r)G^{-1}(\alpha_1, \ldots, \alpha_r) = E. \tag{2.1.10}$$

To the inverse transformation $G^{-1}(\alpha_1, \ldots, \alpha_r)$ there correspond the n following relations, which are the inverses of the relations (2.1.5)

$$x_i = f_i^{-1}(x_1{}', \ldots, x_n{}'). \tag{2.1.11}$$

The possibility of solving relations (2.1.5) with respect to the x_i implies that the Jacobian of the transformation, that is to say, the determinant $|\partial f_i / \partial x_j|$, is not zero. The transformation groups satisfying the previous requirements have been studied by Sophus Lie and belong to the class of LIE GROUPS.

Transposition of the Properties of Finite Groups

When the domain of variation of the parameters allowing us to run over the whole continuous group is restricted, the group is then compact.

As an example of a noncompact group let us cite the group of translations, the elements of which are vectors $\boldsymbol{t}$ of arbitrary translation. The linear representations of these groups have different properties from those of compact groups, to which latter we shall confine ourselves in the chapters which follow.

The ideas contained in the majority of the definitions and properties of finite groups can be transferred to compact groups. Certain proofs use a summation over all elements, finite in number, of the group. An analogous operation exists for compact groups, but instead of a finite sum an integral is used.

2.1.3 Linear Representations

Let us recall the notion given in the Appendix of the representation of a linear group by applying it to the rotation group $\boldsymbol{SO}(3)$.

Rotation Operators

A geometric rotation R is a one-to-one transformation of three-dimensional space. This transformation preserves a point in the space, distances, and angles, as well as the orientation of the trihedra. A point $\boldsymbol{r}'(x', y', z')$ of the space corresponds under a rotation R to the point $\boldsymbol{r}(x, y, z)$, and may be written as

$$\boldsymbol{r}'(x', y', z') = R\boldsymbol{r}(x, y, z). \tag{2.1.12}$$

Let us consider a vector $\boldsymbol{X}(\boldsymbol{r})$ belonging to a vector space V_n of arbitrary dimension n. These vectors can be, for example, functions, tensors, spinors, etc. By definition the operator $\Gamma(R)$ associated with the rotation R is a mathematical entity which, when acting on a vector $\boldsymbol{X}(\boldsymbol{r})$, transforms it into a vector $\boldsymbol{X}(\boldsymbol{r}')$, that is to say

$$\Gamma(R)\boldsymbol{X}(\boldsymbol{r}) = \boldsymbol{X}(R\boldsymbol{r}) = \boldsymbol{X}(\boldsymbol{r}'). \tag{2.1.13}$$

This relation defines the operator $\Gamma(R)$ which acts in the space of the vectors $\boldsymbol{X}(\boldsymbol{r})$. Let us note that in quantum mechanics the operators associated with rotations are defined a little differently. We set

$$\Gamma(R)\boldsymbol{X}(\boldsymbol{r}) = \boldsymbol{X}(R^{-1}\boldsymbol{r}). \tag{2.1.14}$$

This latter definition necessitates changing every angle of rotation into its opposite, for example α into $-\alpha$, in the formulas obtained from relation (2.1.13). Except where the contrary is stated, we shall use definition (2.1.13) in the remainder.

Since the product of two rotations R_1 and R_2 gives another rotation $R_3 = R_1 R_2$, definition (2.1.13) allows us to show that the operators satisfy the fundamental property

$$\Gamma(R_2 R_1) = \Gamma(R_2)\Gamma(R_1). \tag{2.1.15}$$

In fact, using (2.1.13) we have

$$\begin{aligned} \boldsymbol{X}(R_3\boldsymbol{r}) &= \boldsymbol{X}(R_2 R_1 \boldsymbol{r}) \\ &= \Gamma(R_2 R_1)\boldsymbol{X}(\boldsymbol{r}) \\ &= \boldsymbol{X}(R_2[R_1\boldsymbol{r}]) \\ &= \Gamma(R_2)\boldsymbol{X}(R_1\boldsymbol{r}) \\ &= \Gamma(R_2)\Gamma(R_1)\boldsymbol{X}(\boldsymbol{r}). \end{aligned} \tag{2.1.16}$$

As this latter relation holds for an arbitrary vector $\boldsymbol{X}(\boldsymbol{r})$ the relation (2.1.15) is thus satisfied.

Irreducible and Reducible Representations

The correspondence (2.1.13) thus established between geometric rotations and rotation operators preserves the group's law. We say that the set of operators $\Gamma(R)$ forms a REPRESENTATION *of the rotation group*. The vector space V_n is called the REPRESENTATION SPACE, and the basis of this vector space forms a REPRESENTATION BASIS.

Let us consider a subspace V_1 of the representation space V_n and let us assume that all the elements of V_1 are transformed into one another uniquely under the action of all the rotations of the group. We shall say that V_1 is a subspace *stable under* $\boldsymbol{SO}(3)$. The space V_1 defined a representation Γ_1 of the group. If the space V_1 cannot itself be decomposed into other stable subspaces we shall say that the representation Γ_1 is an IRREDUCIBLE REPRESENTATION of $\boldsymbol{SO}(3)$. In the opposite case the representation is called a REDUCIBLE REPRESENTATION.

Matrix Representation

Let $\boldsymbol{e}_1, \boldsymbol{e}_2, \ldots, \boldsymbol{e}_n$ be basis vectors of a representation space V_n. For an arbitrary rotation the transform of every vector $\boldsymbol{e}_i$ can be decomposed in the basis of V_n, that is,

$$\Gamma(R)\boldsymbol{e}_i = \sum_{j=1}^{n} G_{ji}\boldsymbol{e}_j. \tag{2.1.17}$$

The numbers G_{ji} form the matrix elements of the operator $\Gamma(R)$ with respect to the basis $\{\boldsymbol{e}_i\}$ of the space V_n. With each rotation R we can thus associate a matrix $M(R)$ having as elements the quantity G_{ji}. We shall say that the matrices $M(R)$ form a representation *given in matrix form* of the rotation group. From the fundamental property (2.1.15) it is easily deduced (see the Appendix) that the matrices $M(R)$ satisfy the relation

$$M(R_2R_2) = M(R_2)M(R_1). \tag{2.1.18}$$

The matrix elements $M(R)$ are continuous functions of the parameters on which the rotation depends.

REMARK **2.2.** The previous definitions may be applied to all the continuous groups by replacing the rotation operation with the transformation of the group under consideration. Let us note the following terminology:

— If the vector space of representations of a group is a tensor product space it is said that we have a TENSOR REPRESENTATION of the group;

— If the representation space has spinors as elements we have a SPINOR REPRESENTATION. ■

2.1.4 Infinitesimal Generators

As is done for finite symmetry groups, it has just been seen that we can realize representations of continuous groups. But, moreover, as a result of the continuity of the parameters α_i, *infinitesimal operators* will be defined and, as a consequence, *infinitesimal matrices*. In the course of Chapter 1 we have already calculated the infinitesimal rotation matrix in spinor space, given by (1.3.3). Let us give an example of an infinitesimal operator which will lead us to the notion of *infinitesimal generator*.

The Group of Rotations in a Plane

Let us consider an infinitesimal rotation $R_{\mathrm{d}\alpha}$ through an angle $\mathrm{d}\alpha$ in the neighborhood of $\alpha = 0$. Formulas (2.1.1) then become

$$\begin{aligned} x + \mathrm{d}x &= f_1(x, y; \mathrm{d}\alpha), \\ y + \mathrm{d}y &= f_2(x, y; \mathrm{d}\alpha), \end{aligned} \tag{2.1.19}$$

whence, using (2.1.8),

$$\begin{aligned} \mathrm{d}x &= f_1(x, y; \mathrm{d}\alpha) - f_1(x, y; 0) \\ &= \left[\frac{\partial f_1}{\partial \alpha}\right]_{\alpha=0} \mathrm{d}\alpha, \\ \mathrm{d}y &= f_2(x, y; \mathrm{d}\alpha) - f_2(x, y; 0) \\ &= \left[\frac{\partial f_2}{\partial \alpha}\right]_{\alpha=0} \mathrm{d}\alpha. \end{aligned} \tag{2.1.20}$$

Differentiation of (2.1.1) gives us

$$\mathrm{d}x = -y\mathrm{d}\alpha, \qquad \mathrm{d}y = x\mathrm{d}\alpha. \tag{2.1.21}$$

Let us consider for the present a function $F(x, y)$ which is differentiable with respect to x and y. If the rotation R_α transforms the point $M(x, y)$ into $M'(x', y')$, the rotation operator $\Gamma(R_\alpha)$ transforms $F(x, y)$ into $F(x', y')$, that is,

$$F(x', y') = \Gamma(R_\alpha) F(x, y). \tag{2.1.22}$$

An infinitesimal rotation $R_{\mathrm{d}\alpha}$ in the neighborhood of $\alpha = 0$ then gives us

$$\begin{aligned} F(x + \mathrm{d}x, y + \mathrm{d}y) &= \Gamma(R_{\mathrm{d}\alpha}) F(x, y) \\ &= F(x, y) + \frac{\partial F}{\partial x}\mathrm{d}x + \frac{\partial F}{\partial y}\mathrm{d}y. \end{aligned} \tag{2.1.23}$$

Let us replace $\mathrm{d}x$ and $\mathrm{d}y$ by their respective expressions given by (2.1.21). This yields

$$\begin{aligned}\Gamma(R_{\mathrm{d}\alpha})F(x,y) &= F(x,y) + \frac{\partial F}{\partial x}(-y\mathrm{d}\alpha) + \frac{\partial F}{\partial y}(x\mathrm{d}\alpha)\\ &= \left[1 - y\frac{\partial}{\partial x}\,\mathrm{d}\alpha + x\frac{\partial}{\partial y}\,\mathrm{d}\alpha\right]F(x,y). \qquad (2.1.24)\end{aligned}$$

Let us set

$$X = -y\frac{\partial}{\partial x} + x\frac{\partial}{\partial y}\,.$$

The expression (2.1.24) becomes

$$\Gamma(R_{\mathrm{d}\alpha})F(x,y) = [1 + X\mathrm{d}\alpha]F(x,y), \qquad (2.1.25)$$

whence the expression for the infinitesimal rotation operator

$$\Gamma(R_{\mathrm{d}\alpha}) = 1 + X\mathrm{d}\alpha. \qquad (2.1.26)$$

The operator X is called the INFINITESIMAL GENERATOR of the group of plane rotations. This name comes from every rotation R_α being able to be expressed as a function of X, as will be seen in the sequel.

Continuous Groups

Let us now consider an arbitrary continuous group with elements $G(\alpha_1, \ldots, \alpha_r)$ and let $F(x_1, \ldots, x_n)$ be a function which is differentiable with respect to the x_k. An operation $G(\alpha_1, \ldots, \alpha_r)$ transforms the variables x_i into $x_i{}'$ in accordance with relation (2.1.5), and it defines the operators $\Gamma[G(\alpha_1, \ldots, \alpha_r)]$ by

$$F(x_1{}', \ldots, x_n{}') = \Gamma[G(\alpha_1, \ldots, \alpha_r)]F(x_1, \ldots, x_n). \qquad (2.1.27)$$

It will be said that this transformation acting on $F(x_1, \ldots, x_n)$ is INDUCED by G over $(x_1, \ldots, x_n)$.

For the moment let us consider a transformation only a little different from the identity operation E and which corresponds to the values $\alpha_i = 0$. To do so let us take infinitely small values $\mathrm{d}\alpha_j$ of the parameters. The infinitesimal transformation $G(\mathrm{d}\alpha_1, \ldots, \mathrm{d}\alpha_r)$ changes the point $M(x_1, \ldots, x_n)$ into a neighboring point $M(x_1 + \mathrm{d}x_1, \ldots, x_n + \mathrm{d}x_n)$, and by relation (2.1.5) we have

$$x_i + \mathrm{d}x_i = f_i(x_1, \ldots, x_n; \mathrm{d}\alpha_1, \ldots, \mathrm{d}\alpha_r), \qquad (2.1.28)$$

whence, using (2.1.8)

$$\begin{aligned}\mathrm{d}x_i &= f_i(x_1, \ldots, x_n; \mathrm{d}\alpha_1, \ldots, \mathrm{d}\alpha_r) - f_i(x_1, \ldots, x_n; 0, \ldots, 0)\\ &= \sum_{j=1}^{r}\left[\frac{\partial f_i(x_1, \ldots, x_n; \mathrm{d}\alpha_1, \ldots, \mathrm{d}\alpha_r)}{\partial \partial\alpha_j}\right]_{\alpha_j=0}\mathrm{d}\alpha_j. \qquad (2.1.29)\end{aligned}$$

The relation (2.1.27) may then be written for an infinitesimal transformation $G_{\mathrm{d}\alpha} = G(\mathrm{d}\alpha_1, \ldots, \mathrm{d}\alpha_r)$ as

$$\begin{aligned} \Gamma(G_{\mathrm{d}\alpha})F(x_1, \ldots, x_n) &= F(x_1 + \mathrm{d}x, \ldots, x_n + \mathrm{d}x_n) \\ &= F(x_1, \ldots, x_n) + \sum_{i=1}^{n} \frac{\partial F}{\partial x_i} \mathrm{d}x_i. \end{aligned} \tag{2.1.30}$$

Let us replace $\mathrm{d}x_i$ by its expression (2.1.29), obtaining

$$\Gamma(G_{\mathrm{d}\alpha})F(x_1, \ldots, x_n) = \left[1 + \sum_{i=1}^{n} \sum_{j=1}^{r} \left[\frac{\partial f_i}{\partial \alpha_j}\right]_{\alpha_j=0} \frac{\partial}{\partial x_i} \mathrm{d}\alpha_j\right] F(x_1, \ldots, x_n). \tag{2.1.31}$$

Let us set

$$X_j = \sum_{i=1}^{n} \left[\frac{\partial f_i}{\partial \alpha_j}\right]_{\alpha_j=0} \frac{\partial}{\partial x_i}. \tag{2.1.32}$$

The infinitesimal operator $\Gamma(G_{\mathrm{d}\alpha})$ may then be written

$$\Gamma(G_{\mathrm{d}\alpha}) = 1 + \sum_{j=1}^{r} X_j \mathrm{d}\alpha_j. \tag{2.1.33}$$

The infinitesimal operators X_j are called the *infinitesimal generators* of the group. Each corresponds to a parameter α_j. Sophus Lie proved the following theorem:

THEOREM **2.1.** *A necessary and sufficient condition for continuous transformations defined by* (2.1.5) *to form a group is that the commutator of an arbitrary two infinitesimal generators be a linear combination of the generators, that is*

$$X_i X_j - X_j X_i = \sum_{k} \xi_{ijk} X_k. \tag{2.1.34}$$

The coefficients ξ_{ijk} *are called the* STRUCTURE CONSTANTS *of the group and are an essential characteristic of the group.* ■

2.1.5 Infinitesimal Matrices

The operators which represent the transformations of a continuous group can be expressed in a matrix form, and we shall use principally these matrices as representations. The *infinitesimal matrices* are going to correspond to the infinitesimal generators.

Let us consider a continuous group G represented by matrices $M(G)$, and let us write $M_{ij}(G)$ for the matrix elements, which are functions of the r parameters of the group, that is to say, $M_{ij}[G(\alpha_1, \ldots, \alpha_r)]$. Let us define the following matrix elements by differentiating the M_{ij} with respect to the parameters α_k at the point $\alpha_k = 0$ $(k = 1, 2, \ldots, r)$, that is,

$$X^{(k)}{}_{ij} = \left[\frac{\partial M_{ij}}{\partial \alpha_k}\right]_{\alpha_k=0}. \tag{2.1.35}$$

The matrices denoted $X^{(k)}$ which have the quantities $X^{(k)}{}_{ij}$ as their matrix elements are called the *infinitesimal matrices* of the group $\boldsymbol{G}$. The matrix elements M_{ij} can be expanded in a Taylor series in the neighborhood of $\alpha_1 = \alpha_2 = \cdots = \alpha_r = 0$. Restricting ourselves to the first-order terms we obtain

$$M_{ij} = (M_{ij})_0 + \sum_{k=1}^{r} \alpha_k X^{(k)}{}_{ij}, \tag{2.1.36}$$

with

$$(M_{ij})_0 = M_{ij}[G(0, \ldots, 0)],$$

where $G(0, \ldots, 0) = E$ is the identity transformation. Let us write as I the identity matrix which has $(M_{ij})_0$ as its elements. Relation (2.1.28) then gives as the expression for the matrix $M(G)$,

$$M(G) = I + \sum_{k=1}^{r} \alpha_k X^{(k)}. \tag{2.1.37}$$

We obtain a restricted expansion of the matrices of the group in the neighborhood of the unit element of this group. Comparing with relation (2.1.33) it is seen that the matrices $X^{(k)}$ are the matrices of the infinitesimal generators X_k. The infinitesimal matrices satisfy the same commutation relations (2.1.34) as the generators.

Change of Parameters

The parameters of a continuous group are partly arbitrary. For example, the three-dimensional rotations can be expressed as functions of the rotations about the fixed Cartesian axes or as a function of the Euler angles. In general we can substitute for the α_i some parameters β_j such that

$$\beta_j = \beta_j(\alpha_1, \alpha_2, \ldots, \alpha_r), \tag{2.1.38}$$

with $j = 1, \ldots, r$ and a nonzero Jacobian of the transformation. Let us write the infinitesimal matrices of the group as $B^{(j)}$, defined with respect to the parameters β_j. The matrix elements of $B^{(j)}$ are given by

$$\begin{aligned} {B^{(j)}}_{kl} &= \left[\frac{\partial M_{kl}}{\partial \beta_j}\right]_0 \\ &= \sum_{j=1}^{r} {X^{(i)}}_{kl} \left[\frac{\partial \alpha_i}{\partial \beta_j}\right]_0, \end{aligned} \tag{2.1.39}$$

and the infinitesimal matrices may be written

$$B^{(j)} = \sum_{i=1}^{r} X^{(i)} \left[\frac{\partial \alpha_i}{\partial \beta_j}\right]_0. \tag{2.1.40}$$

Hermitian Infinitesimal Matrices

Instead of the infinitesimal matrices $X^{(k)}(G)$ let us introduce the matrices

$$B^{(k)}(G) = iX^{(k)}(G).$$

If the representation is unitary let us show that the matrices $B^{(k)}(G)$ are Hermitian.

To do so relation (2.1.37) gives us

$$\begin{aligned} M(G) &= I + \sum_{k=1}^{r} \alpha_k X^{(k)}(G) \\ &= I - i \sum_{k=1}^{r} \alpha_k B^{(k)}(G), \end{aligned} \tag{2.1.41}$$

and the unitarity condition on the matrices of the representation may be written

$$\begin{aligned} I &= M^\dagger(G) M(G), \\ &= \left[I + i \sum_k \alpha_k B^{(k)\dagger}(G)\right] \left[I - i \sum_k \alpha_k B^{(k)}(G)\right], \end{aligned} \tag{2.1.42}$$

whence by restricting ourselves to the terms of first order in the parameters α_k of the group we obtain

$$\sum_k \alpha_k (B^{(k)\dagger}(G) - B^{(k)}(G)) = 0. \tag{2.1.43}$$

As the parameters α_k are independent we obtain

$$B^{(k)\dagger}(G) = B^{(k)}(G), \tag{2.1.44}$$

which proves the Hermiticity of the matrices $B^{(k)}(G)$.

2.1.6 Exponential Mapping

In the theory of continuous groups it is proved that each Lie group has one-parameter subgroups. Every element of a Lie group can thus be either an element of such a subgroup or a product of such elements. On the other hand, it is proved that each one-parameter subgroup is isomorphic to the set $\mathbb{R}$ of real numbers, or the circle, provided with the law of addition. This permits us to choose as this subgroup a canonical parameter linked in a natural way to the multiplication law of the subgroup, and to determine its elements explicitly as a function of the infinitesimal matrices.

One-Parameter Subgroup

Let us consider a group of matrices $M[G(\alpha_1, \dots, \alpha_r)]$ and a one-parameter subgroup with parameter θ. The elements of the latter are matrices of which all the parameters are differentiable functions of the parameter θ, that is to say, $\alpha_i = \alpha_i(\theta)$. The parameter θ can be chosen so that we have the equalities

$$\begin{aligned} M[G(0)] &= I, \\ M[G(\theta_1)]M[G(\theta_2)] &= M[G(\theta_1 + \theta_2)]. \end{aligned} \tag{2.1.45}$$

Let us differentiate the matrix product with respect to θ_1. We obtain

$$M[G(\theta_2)]\frac{\mathrm{d}M[G(\theta_1)]}{\mathrm{d}\theta_1} = \frac{\mathrm{d}M[G(\theta_1 + \theta_2)]}{\mathrm{d}\theta_1}. \tag{2.1.46}$$

Let us set

$$X^\theta = \left[\frac{\mathrm{d}M[G(\theta)]}{\mathrm{d}\theta}\right]_{\theta=0}.$$

For $\theta_1 = 0$ and $\theta_2 = 0$ relation (2.1.46) may be written

$$M[G(\theta)]X^\theta = \frac{\mathrm{d}M[G(\theta)]}{\mathrm{d}\theta}. \tag{2.1.47}$$

The solution of this matrix differential equation is obtained for the condition $M[G(0)] = I$ and we have

$$M[G(\theta)] = \exp(X^\theta \theta), \tag{2.1.48}$$

where $\exp[X^\theta \theta]$ is the matrix defined by the series

$$\exp[X^\theta \theta] = I + X^\theta \theta + \frac{1}{2!}(X^\theta \theta)^2 + \frac{1}{3!}(X^\theta \theta)^3 + \cdots . \tag{2.1.49}$$

The matrix X^θ is obtained as a function of the infinitesimal matrices $X^{(k)}$ in the parameter α_k. Using relation (2.1.40), that is,

$$X^\theta = \sum_{k=1}^{r} X^{(k)} \left[\frac{\mathrm{d}\alpha_k}{\mathrm{d}\theta}\right]_{\theta=0}, \tag{2.1.50}$$

and substituting (2.1.50) into (2.1.48) we obtain

$$M[G(\theta)] = \exp\left[\sum_{k=1}^{r} X^{(k)} \left[\frac{\mathrm{d}\alpha_k}{\mathrm{d}\theta}\right]_{\theta=0} \theta\right]. \qquad (2.1.51)$$

2.1.7 The Nomenclature of Continuous Linear Groups

Let us give some examples of familiar groups in considering quite generally complex parameters and transformations operating in a space of complex variables. Therefore we shall not state in the notations which follow that the groups are defined over the complex numbers.

Linear Groups

Let us consider the transformation which brings (z_1', z_2') into correspondence with the pair of complex variables (z_1, z_2) by

$$z_1' = a_1 z_1 + a_2 z_2, \qquad z_2' = a_3 z_1 + a_4 z_2 \qquad (2.1.52)$$

where the parameters a_i are also complex. As the real and imaginary parts of these parameters are independent we have a transformation with eight real parameters. By considering all the values of the continuous parameters a_i such that the determinant of the system (2.1.52) is different from zero, we obtain the transformations (2.1.52) which form the two-dimensional *continuous linear group* denoted $\boldsymbol{GL}(2)$. It is isomorphic to the group of invertible matrices of order 2 which have the parameters a_i as their matrix elements.

The n-dimensional linear transformations of the form

$$z_1' = \sum_j a_{ij} z_j, \qquad i = 1, \ldots, n, \qquad (2.1.53)$$

with $|a_{ij}| \neq 0$, form the n-dimensional linear group denoted $\boldsymbol{GL}(n)$. It is a group with $2n^2$ real parameters. It is isomorphic to the group of invertible square matrices of order n. To state whether they are over the reals or the complex numbers we use the notations $\boldsymbol{GL}(n, \mathbb{R})$ or $\boldsymbol{GL}(n, \mathbb{C})$.

Unitary Groups

The unitary $\boldsymbol{U}(n)$ is isomorphic to the group of unitary square matrices of order n. As the elements of unitary matrices are related with each other by n^2 conditions of orthogonality and normalization, the unitary group $\boldsymbol{U}(n)$ is a group with $(2n^2 - n^2) = n^2$ independent parameters. Since we have the unitarity condition $\sum |a_{ij}|^2 = 1$ we see that $|a_{ij}|^2 \leqslant 1$ for every value of i and j. The group $\boldsymbol{U}(n)$ is a compact group and all the subgroups of $\boldsymbol{U}(n)$ are compact.

The unitary unimodular group $\boldsymbol{SU}(n)$ is the subgroup of $\boldsymbol{U}(n)$ obtained by restriction to the square matrices with determinant equal to unity. The number of its parameters is equal to $n^2 - 1$.

Orthogonal Groups

The orthogonal group $\boldsymbol{O}(n)$ is isomorphic to the group of orthogonal square matrices or order n. The conditions of normalization and orthogonality of the matrix elements impose $n + \frac{1}{2}n(n-1)$ relations, which leaves the number of independent parameters equal to $\frac{1}{2}n(n-1)$. The determinant of an orthogonal matrix is equal to $+1$ or -1.

The rotation group $\boldsymbol{SO}(n)$ is isomorphic to the group of orthogonal square matrices of order n with determinant equal to $+1$. This is the group of three-dimensional rotations $\boldsymbol{SO}(3, \mathbb{R})$, which is more interesting for applications in physics. It is a group with $\frac{1}{2}3(3-1) = 3$ parameters, the transformations of which leave the distance from one point to another invariant.

All these continuous groups are particular cases of *Lie groups*. We have only studied some properties of them which are useful for the applications we have in view.

2.2 Unimodular Unitary Groups

In the course of the previous chapter we have seen that the two-component spinors are able to form a two-dimensional vector space over $\mathbb{C}$. This latter is going to form a representation space for the unimodular unitary group. The n-component spinors will form the representation spaces of arbitrary dimension of this group.

2.2.1 The Unitary Group $\boldsymbol{U}(2)$

Let us consider the linear transformation U which brings the pair (ψ', ϕ') into correspondence with the pair of complex variables (ψ, ϕ) by

$$\psi' = a\psi + b\phi, \qquad \phi' = c\psi + d\phi, \tag{2.2.1}$$

where the parameters a, b, c, d are also complex. If these parameters are continuous and if the determinant of the system (2.2.1) is different from zero, these transformations U form the two-dimensional linear group denoted $\boldsymbol{GL}(2)$.

Let us consider the pairs (ψ, ϕ) as belonging to a vector space V_2, and let us take for the basis of V_2 the natural basis $\boldsymbol{e}_1 = (1, 0)$ and $\boldsymbol{e}_2 = (0, 1)$. The vector $\chi = (\psi, \phi)$ then has as its components the numbers (ψ, ϕ) themselves, and the matrix $M(U)$ of the representation of the group $\boldsymbol{GL}(2)$

may be written

$$M(U) = \begin{bmatrix} a & b \\ c & d \end{bmatrix}. \tag{2.2.2}$$

Unitary Matrices

Let $\boldsymbol{x}$ and $\boldsymbol{y}$ be vectors of a vector space V_n. Let us write $\Gamma(U)\boldsymbol{x}$ and $\Gamma(U)\boldsymbol{y}$, respectively, for the transforms of the vectors $\boldsymbol{x}$ and $\boldsymbol{y}$ under the operator representing the transformation U. This transformation is said to be unitary if it preserves the Hermitian product, that is to say

$$\boldsymbol{x}\cdot\boldsymbol{y} = \Gamma(U)\boldsymbol{x}\cdot\Gamma(U)\boldsymbol{y}.$$

The operators are said to be unitary as well as the matrices $M(U)$ which represent them in a given basis. We then prove that the unitary matrices are such that

$$\begin{aligned} M^\dagger(U)M(U) &= M(U)M^\dagger(U) \\ &= I, \end{aligned}$$

where $M^\dagger(U)$ is the adjoint matrix of $M(U)$, and I is the unit matrix. For the moment let us consider the matrix (2.2.2). In order that it be unitary we must have

$$\begin{aligned} \begin{bmatrix} a & b \\ c & d \end{bmatrix} \begin{bmatrix} a^* & c^* \\ b^* & d^* \end{bmatrix} &= \begin{bmatrix} |a|^2 + |b|^2 & ac^* + bd^* \\ ca^* + db^* & |c|^2 + |d|^2 \end{bmatrix} \\ &= \begin{bmatrix} 1 & 0 \\ 0 & 1 \end{bmatrix}, \end{aligned}$$

that is to say,

$$\begin{aligned} |a|^2 + |b|^2 &= 1, \\ |c|^2 + |d|^2 &= 1, \\ d &= -c\left(\frac{a^*}{b^*}\right). \end{aligned} \tag{2.2.3}$$

These three relations gives us $|c| = |b|$, which leaves four independent real parameters, as can be seen by writing

$$a = e^{iu}\cos\lambda, \qquad b = e^{iv}\sin\lambda, \qquad c = -e^{iw}\sin\lambda. \tag{2.2.4}$$

The relations (2.2.3) are then satisfied. The set of these unitary transformations form a group, as can be verified from the properties. It is the unitary group $\boldsymbol{U}(2)$. With the parameters (2.2.4), the matrices $M(U)$ of the transformations of $\boldsymbol{U}(2)$ may be written

$$M(U) = \begin{bmatrix} e^{iu}\cos\lambda & e^{iv}\sin\lambda \\ -e^{iw}\sin\lambda & e^{i(w+v-u)}\cos\lambda \end{bmatrix}, \tag{2.2.5}$$

and its determinant has the value

$$\begin{aligned} \Delta &= e^{i(v+w)}(\cos^2\lambda + \sin^2\lambda) \\ &= e^{i(v+w)}, \end{aligned} \tag{2.2.6}$$

whence

$$|\Delta| = 1.$$

2.2.2 *The Unitary Unimodular Group* $\boldsymbol{SU}(2)$

If the additional condition $\Delta = 1$ is imposed we have $v = -w$ (to within $2k\pi$) and a subgroup of $\boldsymbol{U}(2)$ is obtained. In fact, if T and T' are two transformations of $\boldsymbol{U}(2)$ such that

$$\Delta(T) = 1 \qquad \text{and} \qquad \Delta(T') = 1,$$

then

$$\Delta(TT') = \Delta(T)\Delta(T') = 1.$$

For $v = -w$ the parameters (2.2.4) become $b = -c^*$, $d = a^*$, and the unitary transformation (2.2.1), when it is unimodular, becomes

$$\begin{aligned} \psi' &= a\psi + b\phi, \\ \phi' &= -b^*\psi + a^*\phi, \end{aligned} \tag{2.2.7}$$

with

$$aa^* + bb^* = 1.$$

The unimodular subgroup is called a UNIMODULAR UNITARY SUBGROUP, and is denoted $\boldsymbol{SU}(2)$. It depends upon no more than three real parameters, u, v, λ, like the group $\boldsymbol{SO}(3)$ of spacial rotations.

As the matrix representation of the group $\boldsymbol{SU}(2)$ has as its representation space the vector space formed from the vectors (ψ, ϕ), it has as its matrices

$$M(U) = \begin{bmatrix} a & b \\ -b^* & a^* \end{bmatrix}. \tag{2.2.8}$$

Two-Component Spinors

We may recover the matrices (1.2.14) from Chapter 1 which define the two-component spinor rotations. We have seen that these rotations in spinor space are associated with rotations in three-dimensional space. Thus to every transformation of the group $\boldsymbol{SU}(2)$ there corresponds a rotation belonging to the group $\boldsymbol{SO}(3)$.

The converse must be modified a little, as is going to be seen. In fact, if a is changed into b and $-a$ into $-b$, then ψ' and ϕ' change sign under the transformation (2.2.7). However, to the pair (ψ, ϕ) there correspond the coordinates x, y, z given by (1.1.6) and (1.1.9), where there appear the products of ψ and ϕ with the conjugates. The transforms x', y', z' corresponding to the pair (ψ', ϕ') thus remain invariant under a change of sign of a and b. Consequently, to every rotation R of the group $\boldsymbol{SO}(3)$ there correspond *two* transformations U and $-U$ of the unimodular unitary group.

The Infinitesimal Rotation Matrix

We have seen that for a rotation in three-dimensional space characterized by an angle α and a unitary vector $\boldsymbol{L} = (L_1, L_2, L_3)$ along the axis of rotation, the matrix of the rotation in spinor space is given by expression (1.3.1). Definition (2.1.27) of the elements of an infinitesimal matrix of a continuous group gives us for the infinitesimal matrix of the group $\boldsymbol{SU}(2)$

$$X(\boldsymbol{L}) = \begin{bmatrix} -iL_3 & -(iL_1 + L_2) \\ (-iL_1 + L_2) & iL_3 \end{bmatrix}. \tag{2.2.9}$$

The Pauli matrices $\sigma_1, \sigma_2, \sigma_3$ given by (1.3.12) allow us to write the infinitesimal matrix in the form

$$\begin{aligned} X(\boldsymbol{L}) &= -i(L_1\sigma_1 + L_2\sigma_2 + L_3\sigma_3) \\ &= -i\boldsymbol{L}\cdot\boldsymbol{\sigma}, \end{aligned} \tag{2.2.10}$$

where $\boldsymbol{\sigma}$ is the vector which has the Pauli matrices as its components

$$\boldsymbol{\sigma} = (\sigma_1, \sigma_2, \sigma_3).$$

The Exponential Expression for the Rotation Matrix

The rotation matrices in the space of spinors may be written in the form (1.3.17), that is to say,

$$M[U(\boldsymbol{L}, \tfrac{1}{2}\alpha)] = \cos\tfrac{1}{2}\alpha\sigma_0 + i\sin\tfrac{1}{2}\alpha\boldsymbol{L}\cdot\boldsymbol{\sigma}. \tag{2.2.11}$$

Since $\boldsymbol{L}$ is a unitary vector we have

$$(\boldsymbol{L}\cdot\boldsymbol{\sigma})^2 = \sigma_0. \tag{2.2.12}$$

Let us consider the following series expansion of the matrix operator

$$\begin{aligned} e^{i\boldsymbol{L}\cdot\boldsymbol{\sigma}\alpha/2} &= \sigma_0 - i\boldsymbol{L}\cdot\boldsymbol{\sigma}(\tfrac{1}{2}\alpha) - \frac{(\boldsymbol{L}\cdot\boldsymbol{\sigma})^2}{2!}(\tfrac{1}{2}\alpha)^2 + \frac{(\boldsymbol{L}\cdot\boldsymbol{\sigma})^3}{3!}(\tfrac{1}{2}\alpha)^3 + \cdots \\ &= \sigma_0\left(1 - \frac{1}{2!}(\tfrac{1}{2}\alpha)^2 + \cdots\right) - i\boldsymbol{L}\cdot\boldsymbol{\sigma}\left(\tfrac{1}{2}\alpha - \frac{1}{3!}(\tfrac{1}{2}\alpha)^3\right). \end{aligned} \tag{2.2.13}$$

This latter expression introduces the expansion of $\cos(\frac{1}{2}\alpha)$ and $\sin(\frac{1}{2}\alpha)$, whence

$$e^{-i\boldsymbol{L}\cdot\boldsymbol{\sigma}\alpha/2} = \sigma_0 \cos\tfrac{1}{2}\alpha - i\boldsymbol{L}\cdot\boldsymbol{\sigma}\sin\tfrac{1}{2}\alpha. \tag{2.2.14}$$

Comparing expressions (2.2.11) and (2.2.14) we obtain the following remarkable form for the rotation matrices of the group $\boldsymbol{SU}(2)$

$$\begin{aligned} M[U(\boldsymbol{L}, \tfrac{1}{2}\alpha)] &= e^{i\boldsymbol{L}\cdot\boldsymbol{\sigma}\alpha/2} \\ &= e^{X(\boldsymbol{L})\alpha/2}. \end{aligned} \tag{2.2.15}$$

We obtain the general exponential form (2.1.48) which expresses in exponential form the matrices of the group as functions of the infinitesimal matrix, which for $\boldsymbol{SU}(2)$ is the matrix $X(\boldsymbol{L})$.

2.2.3 Three-Dimensional Representations

The unimodular unitary group has as its two-dimensional irreducible representation the matrices (2.2.8) of which the spinors (ψ, ϕ) form a representation space. We can form a three-dimensional irreducible representation by considering the following monomials

$$\zeta_0 = \frac{1}{\sqrt{2}}\psi^2, \qquad \zeta_1 = \psi\phi, \qquad \zeta_2 = \frac{1}{\sqrt{2}}\phi^2, \tag{2.2.16}$$

as the components of a vector. The coefficients $1/\sqrt{2}$ are introduced so as to preserve the Hermitian product under the unitary transformation (2.2.7). Let $\boldsymbol{x} = (\zeta_0, \zeta_1, \zeta_2)$, then the Hermitian product of $\boldsymbol{x}$ with itself is

$$\tfrac{1}{2}(\psi\psi^*)^2 + \psi\psi^*\phi\phi^* + \tfrac{1}{2}(\phi\phi^*)^2 = \tfrac{1}{2}(\psi\psi^* + \phi\phi^*)^2. \tag{2.2.17}$$

If the spinor (ψ, ϕ) is unitary we have

$$(\psi\psi^* + \phi\phi^*) = 1,$$

and the Hermitian product of the vector $(\zeta_0, \zeta_1, \zeta_2)$ is equal to a constant. These vectors form a three-dimensional vector space and are able to serve as a representation space for the group $\boldsymbol{SU}(2)$.

Let us submit the quantities ψ, ϕ to the unitary transformation (2.2.7) and let us then calculate the transformations undergone by the monomials (2.2.16). We obtain

$$\begin{aligned}
\zeta_0{}' &= \frac{1}{\sqrt{2}}(a\psi + b\phi)^2 \\
&= a^2\zeta_0 + \sqrt{2}ab\zeta_1 + b^2\zeta_2, \\
\zeta_1{}' &= (a\psi + b\phi)(-b^*\psi + a^*\phi) \\
&= -\sqrt{2}ab^*\zeta_0 + (aa^* - bb^*)\zeta_1 + \sqrt{2}a^*b\zeta_2, \\
\zeta_2{}' &= \frac{1}{\sqrt{2}}(-b^*\psi + a^*\phi)^2 \\
&= b^{*2}\zeta_0 - \sqrt{2}a^*b^*\zeta_1 + a^{*2}\zeta_2.
\end{aligned} \tag{2.2.18}$$

The transformations undergone by $\zeta_0, \zeta_1, \zeta_2$ again give the elements of the initial vector space. The matrix of this transformation is

$$M^{(1)}(U) = \begin{bmatrix} a^2 & \sqrt{2}ab & b^2 \\ -\sqrt{2}ab^* & aa^* - bb^* & \sqrt{2}a^*b \\ b^{*2} & -\sqrt{2}a^*b^* & a^{*2} \end{bmatrix}. \tag{2.2.19}$$

The vector with the components $\zeta_0, \zeta_1, \zeta_2$, and which transform under the matrix (2.2.19), are second-order spinors.

*2.2.4 Representations of the Groups **SU**(2)*

Spinors with $2j+1$ Components

We can form representations of higher dimension by considering the following $n+1$ monomials of degree n

$$\psi^n,\ \psi^{n-1}\phi,\ \ldots,\ \psi\phi^{n-1},\ \phi^n \tag{2.2.20}$$

as the components of a vector belonging to an $(n+1)$-dimensional vector space V_{n+1}. This latter is able to serve as the representation space for $\boldsymbol{SU}(2)$. In fact, let us subject ψ and ϕ to a transformation U defined by (2.2.1). We obtain for the transform of an arbitrary monomial

$$\begin{aligned}
\psi'^{n-p}\phi'^{p} &= (a\psi + b\phi)^{n-p}(-b^*\psi + a^*\phi)^p \\
&= \sum_{k=0}^{n} U_{pk}{}^{(n)}\psi^{n-k}\phi^k.
\end{aligned} \tag{2.2.21}$$

The $n+1$ monomials (2.2.20) all undergo between themselves a linear transformation by the matrix $M^{(n)}(U) = [U_{pk}{}^{(n)}]$. This latter equality can be obtained because all the monomials (2.2.20) are of the same degree and the transformation U again gives us vectors of the space V_{n+1}. The monomials (2.2.20) are the components of the vectors of the representation space V_{n+1}. These vectors are transformed by rotations according to the relation (2.2.21) and are $(n+1)$-component *spinors*.

The transformation (2.2.21) is not unitary, but it becomes so if instead of (2.2.21) we take the monomials

$$q_k = \frac{\psi^{n-k}\phi^k}{\sqrt{(n-k)!k!}}, \tag{2.2.22}$$

for, taking account of the unitarity condition $\psi\psi^* + \phi\phi^* = 1$ we obtain

$$\begin{aligned} \sum_{k=0}^{n} q_k^* q_k &= \sum_{k=0}^{n} \frac{(\psi\psi^*)^{n-k}(\phi\phi^*)^k}{(n-k)!k!} \\ &= \frac{1}{n!}(\psi\psi^* + \phi\phi^*) \\ &= \text{const.} \end{aligned} \tag{2.2.23}$$

The Hermitian product is therefore preserved under the unitary transformation U since $\psi^*\psi^* + \phi\phi^*$ is invariant, and the same is true for all the powers of this invariant. In order to take account of the customary quantum notations we write

$$n = 2j, \qquad m = j - k. \tag{2.2.24}$$

The variables q_k given by (2.2.22) then take the symmetric form

$$q_m{}^{(j)} = \frac{\psi^{j+m}\phi^{j-m}}{\sqrt{(j+m)!\,(j-m)!}}, \qquad m = -j, \ldots, j-1, j, \tag{2.2.25}$$

with $j = \frac{1}{2}, 1, \frac{3}{2}, 2, \ldots$ since n is the sequence of integers. The number j is called the WEIGHT *of the representation* of which the dimension is $2j+1$. This representation will be denoted $\Gamma^{(j)}$. The $(2j+1)$-component spinors are called j-SPINORS, or, again, SPINORS OF ORDER $2j$.

The Matrices of a Representation

In order to find the matrices of the representation it is necessary to apply the transformation $U(a,b)$, that is,

$$U(a,b)q_m{}^{(j)} = \frac{(a\psi + b\phi)^{j+m}(-b^*\psi + a^*\phi)^{j-m}}{\sqrt{(j+m)!\,(j-m)!}}. \tag{2.2.26}$$

The expansion of this expression with the binomial formula leads to the following general expression for the matrix elements

$$U^{(j)}{}_{m'm} = \sum_{\lambda} \frac{[(j+m)!\,(j-m)!\,(j+m')!\,(j-m')!]^{1/2}}{(j+m-\lambda)!\,(j-m'-\lambda)!\,(m'-m+\lambda)!} \times a^{j+m-\lambda}(a^*)^{j-m'-\lambda} b^{\lambda}(-b^*)^{m'-m+\lambda}. \qquad (2.2.27)$$

The summation over λ is carried out over all the values which actually correspond to a term in the binomial formula, that is,

λ integral, $\quad j+m-\lambda \geqslant 0, \quad j'-m'-\lambda \geqslant 0, \quad \lambda \geqslant 0, \quad m'-m+\lambda \geqslant 0.$

These values correspond to the factorials of the denominator which are not infinite. For example, for $m' = j$ the denominator becomes infinite except for $\lambda = 0$. In fact, the term $(-\lambda)!$, the factorial of a negative integer, is infinite as a result of the relation between the gamma functions and the factorials, $n! = \Gamma(n+1)$. In this case only one term appears in the sum (2.2.27), whence

$$U^{(j)}{}_{jm}(a,b) = \sqrt{\frac{(2j)!}{(j+m)!\,(j-m)!}}\, a^{j+m}(-b^*)^{j-m}. \qquad (2.2.28)$$

2.2.5 Irreducible Representations of $\boldsymbol{SU}(2)$

To do so we are going to show that for a given value of j the j-spinors give a representation $\Gamma^{(j)}$ of $\boldsymbol{SU}(2)$ which is *irreducible*. We are going to use Schur's First Lemma, the proof of which we shall give below.

LEMMA **2.1 (Schur).** *If a matrix A commutes with all the matrices $M(G)$ of an irreducible representation Γ then A is a multiple of the unit matrix.*

Proof. . By hypothesis

$$M(G)A = AM(G). \qquad (2.2.29)$$

Let V_n be the representation space of Γ. In this space there exists at least one eigenvector of the matrix A. Let us write this vector as $\boldsymbol{x}$, which is such that

$$A\boldsymbol{x} = \lambda \boldsymbol{x}, \qquad (2.2.30)$$

where $\boldsymbol{x}$ also represents the matrix of the components of the vector $\boldsymbol{x}$. Applying one of the transformations of the group to $\boldsymbol{x}$ we obtain

$$M(G)\boldsymbol{x} = \boldsymbol{y}. \qquad (2.2.31)$$

Let us show that $\boldsymbol{y}$ is also an eigenvector of A associated with the eigenvalue λ. Equations (2.2.29) and (2.2.30) give us

$$\begin{aligned} A\boldsymbol{y} = AM(G)\boldsymbol{x} &= M(G)A\boldsymbol{x} \\ &= M(G)\lambda\boldsymbol{x} = \lambda\boldsymbol{y}. \end{aligned} \tag{2.2.32}$$

The vector subspace generated by the eigenvectors of A associated with the same eigenvalue is thus stable under G. However, since the representation Γ is assumed to be irreducible this subspace must coincide with V_n completely. As a consequence the matrix A transforming every vector $\boldsymbol{x}$ of V_n into a collinear vector $\lambda\boldsymbol{x}$ is a matrix such that $A = \lambda I_n$, where I_n is the unit matrix of order n. ■

A converse of this theorem can be used as a test of irreducibility of a representation. If it can be proved that the only matrix commuting with all the matrices of a representation of a group is a matrix which is a multiple of the unit matrix, then the representation is irreducible. In fact, in the presence of a true stable subspace (the case of reducibility) it is easy to construct a matrix which commutes and which is not a multiple of the unit matrix.

The Irreducibility of a Representation $\Gamma^{(j)}$

First of all let us consider diagonal matrices $M(U)$ of a representation of $\boldsymbol{SU}(2)$. We then have $b = 0$, $|a| = 1$, and we can write $a = e^{i\alpha/2}$ with arbitrary α. The sole nonzero term in (2.2.22) is then given for $\lambda = 0$ and $m' = m$, and we have

$$\begin{aligned} U^{(j)}{}_{m'm} &= \delta_{m'm}(e^{i\alpha/2})^{j+m}(e^{-i\alpha/2})^{j-m} \\ &= \sigma_{m'm}e^{im\alpha}. \end{aligned} \tag{2.2.33}$$

Let us assume for the moment that there exists a matrix A which commutes with all the matrices $M(U)$, and evidently with all those which are diagonal. By hypothesis $AM(U) = M(U)A$, and for the matrix elements we can write

$$A_{mm'}U_{m'm'} = U_{mm}A_{mm'}, \tag{2.2.34}$$

so for diagonal matrices

$$A_{mm'}(e^{im'\alpha} - e^{im\alpha}) = 0. \tag{2.2.35}$$

As a result $A_{mm'} = 0$ when $m \neq m'$, and the matrix A must be diagonal. For the moment let us consider an arbitrary matrix $M(U)$. The commutation relation with A gives us

$$A_{mm}U_{mm'} = U_{mm'}A_{m'm'}, \tag{2.2.36}$$

and, since $U_{mm'} \neq 0$ by hypothesis, we have $A_{mm} = A_{m'm'}$. The matrix A is therefore not only diagonal but it is also a constant matrix, a multiple of the unit matrix. Consequently, the matrices $M(U)$ form an irreducible representation of the group $\boldsymbol{SU}(2)$.

We prove further that *all* the irreducible representations of $\boldsymbol{SU}(2)$ are obtained from representation spaces formed by the j-spinors, where j runs over all the integer and half-integer values.

2.3 Solved Problems

EXERCISE **2.1**

Consider a function $F(z)$ the expansion of which in an entire series in z is

$$F(z) = \sum_{n=0}^{\infty} f_n z^n.$$

Let A be an operator with which we form the series $\sum_{n=0}^{\infty} F_n A^n$. This series will be denoted by the same symbol as the function F, that is to say, $F(A)$.

1. Let ψ_a be an eigenfunction of A corresponding to the eigenvalue a

$$A\psi_a = a\psi_a.$$

Prove that ψ_a is also an eigenfunction of $F(A)$ and calculate the corresponding eigenvalue.

2. Prove that the matrix of an operator A is diagonal in the orthonormal basis of its eigenfunctions.

3. Consider the following Pauli matrix

$$\sigma_z = \begin{bmatrix} 1 & 0 \\ 0 & -1 \end{bmatrix}.$$

Taking into account the result of Exercise 2, calculate the matrix $\exp(\sigma_z)$.

SOLUTION **2.1**

1. The successive application of A for n times on the eigenfunction ψ_a gives

$$A^n \psi_a = a^n \psi_a,$$

whence

$$\begin{aligned} F(A) &= \sum_{n=0}^{\infty} f_n A^n \psi_a \\ &= \sum_{n=0}^{\infty} f_n a^n \psi_a \\ &= \psi_a \sum_{n=0}^{\infty} f_n a^n. \end{aligned}$$

According to the definition of the function $F(z)$ we have

$$F(a) = \sum_{n=0}^{\infty} f_n a^n,$$

whence

$$F(A)\psi_a = F(a)\psi_a.$$

The function ψ_a is the eigenfunction of the operator $F(A)$ for the eigenvalue $F(a)$.

2. The matrix elements A_{ij} of an operator are given by the Hermitian products

$$A_{ij} = \langle \psi_i | A | \psi_j \rangle .$$

Since $A\psi_j = a_j \psi_j$ this yields

$$\begin{aligned} A_{ij} &= \langle \psi_i | a_j \psi_j \rangle \\ &= a_j \langle \psi_i | \psi_j \rangle \\ &= a_j \delta_{ij}. \end{aligned}$$

Consequently the matrix of the operator A is a diagonal matrix the elements of which located on the principal diagonal are the eigenvalues. We can then define the operator $F(A)$ as the operator which in the basis $\{\psi_n\}$ is represented by the diagonal matrix elements $F(a_n)$.

3. The matrix σ_z has the diagonal elements 1 and -1, which are the eigenvalues of an operator which will be denoted S_z. To these eigenvalues there correspond the eigenfunctions forming an eigen-basis. The operator $\exp(S_z)$ thus has as its diagonal elements of its representative matrix in the eigen-basis the numbers $\exp(1)$ and $\exp(-1)$, whence the matrix

$$\exp(\sigma_z) = \begin{bmatrix} e^1 & 0 \\ 0 & e^{-1} \end{bmatrix} = \begin{bmatrix} e & 0 \\ 0 & \frac{1}{e} \end{bmatrix}.$$

The operator S_z is then, by definition, the operator represented in the eigen-basis by the matrix $\exp(\sigma_z)$.

EXERCISE **2.2**

Let $A(\alpha)$ be an operator which depends on the parameter α. By definition the derivative of $A(\alpha)$ with respect to α is the operator given by the limit under the condition that there exists

$$\frac{\mathrm{d}A}{\mathrm{d}\alpha} = \lim_{\Delta\alpha\to 0} \frac{A(\alpha+\Delta\alpha) - A(\alpha)}{\Delta\alpha}.$$

1. Calculate the derivative of the operator $\exp(\alpha A)$, where A is an operator which does not depend on α.

2. For the moment let $A(\alpha)$ be an operator which depends on α. Calculate the matrix elements of the matrix representing the operator $\mathrm{d}A/\mathrm{d}\alpha$ as a function of the matrix elements A_{ij} of the matrix $M(A)$.

SOLUTION **2.2**

1. The operator $\exp(\alpha A)$ has as its expression

$$e^{\alpha A} = \sum_{n=0}^{\infty} \frac{(\alpha A)^n}{n!}.$$

The derivative of $(\alpha A)^n$ is given by

$$\begin{aligned} \frac{\mathrm{d}(\alpha A)^n}{\mathrm{d}\sigma} &= \lim_{\Delta\alpha\to 0} \frac{[(\alpha+\Delta\alpha)A]^n - (\alpha A)^n}{\Delta\alpha} \\ &= \lim_{\Delta\alpha\to 0} \frac{[(\alpha+\Delta\alpha)^n - \alpha^n]A^n}{\Delta\alpha}. \end{aligned}$$

We obtain the classic derivative of the function α^n which is multiplied by A^n, whence

$$\frac{\mathrm{d}(\alpha A)^n}{\mathrm{d}\alpha} = n\alpha^{n-1}A^n.$$

The derivative of the operator $\exp(\alpha A)$ is obtained by forming the sum of the different derivatives, that is

$$\begin{aligned}\frac{\mathrm{d}e^{\alpha A}}{\mathrm{d}\alpha} &= \sum_{n=1}^{\infty} \frac{n\alpha^{n-1}A^n}{n!} \\ &= A\sum_{p=0}^{\infty} \frac{(\alpha A)^p}{p!} \\ &= Ae^{\alpha A}.\end{aligned}$$

2. The matrix elements of $M(A)$ on an arbitrary basis $\{\psi_j\}$ independent of α are of the form

$$A_{ij} = \langle \psi_i | A | \psi_j \rangle .$$

Let us write $(\mathrm{d}A/\mathrm{d}\alpha)_{ij}$ for the matrix elements of the operator $(\mathrm{d}A/\mathrm{d}\alpha)$. By definition we have

$$\begin{aligned}\left(\frac{\mathrm{d}A}{\mathrm{d}\alpha}\right)_{ij} &= \left\langle \psi_i \left| \frac{\mathrm{d}A}{\mathrm{d}\alpha} \right| \psi_j \right\rangle \\ &= \left\langle \psi_i \left| \frac{\mathrm{d}A\psi_j}{\mathrm{d}\alpha} \right.\right\rangle \\ &= \frac{\mathrm{d}}{\mathrm{d}\alpha} \langle \psi_i | A | \psi_j \rangle \\ &= \frac{\mathrm{d}A_{ij}}{\mathrm{d}\alpha}.\end{aligned}$$

The matrix elements $(\mathrm{d}A/\mathrm{d}\alpha)_{ij}$ of the matrix representing the operator $\mathrm{d}A/\mathrm{d}\alpha$ are thus the derivatives with respect to α of the matrix elements A_{ij} of the matrix representing the operator A.

Exercise **2.3**

Let A be a square matrix of order n and with elements a_{ij}. The matrix elements of the matrix A^k will be denoted ${a_{ij}}^{(k)}$.

1. From the evident majorization $|a_{ij}| \leqslant M$, for any i and j, deduce a majorization for ${a_{ij}}^{(2)}, {a_{ij}}^{(3)}$, and then for ${a_{ij}}^{(k)}$.

2. If the n^2 numerical series

$$S_{ij} = \delta_{ij} + \frac{a_{ij}}{1!} + \frac{{a_{ij}}^{(2)}}{2!} + \cdots + \frac{{a_{ij}}^{(k)}}{k!} + \cdots ,$$

are convergent, it will be said that the matrix series

$$\exp(A) = \mathbb{1} + \frac{A}{1!} + \frac{A^2}{2!} + \cdots + \frac{A^k}{k!} + \cdots$$

converges, and the matrix formed by the sum of this series will be written $\exp(A)$. Using the majorization of the ${a_{ij}}^{(k)}$ show that the series denoted $\exp(A)$ converges for every matrix A.

3. Show that a sufficient condition for

$$\exp(A)\exp(B) = \exp(A+B)$$

to hold is that the matrices A and B commute.

SOLUTION **2.3**

1. The elements of A^2 are given by

$${a_{ij}}^{(2)} = \sum_{k=1}^{n} a_{ik}a_{kj}.$$

The majorization $|a_{ij}| \leqslant M$ implies

$$\begin{aligned} |{a_{ij}}^{(2)}| &\leqslant \sum_{k=1}^{n} |a_{ik}||a_{kj}| \\ &\leqslant nM^2. \end{aligned}$$

Similarly,

$$|{a_{ij}}^{(3)}| \leqslant n^2M^3,$$

whence by induction

$$|{a_{ij}}^{(k)}| \leqslant n^{k-1}M^k.$$

2. The series S_{ij} is majorized in modulus by the series

$$1 + \frac{M}{1!} + \frac{nM^2}{2!} + \cdots + \frac{n^{k-1}M^k}{k!} + \cdots,$$

which can be written

$$\frac{n-1}{n} + \frac{1}{n}\left[1 + \frac{nM}{1!} + \frac{n^2M^2}{2!} + \cdots + \frac{n^kM^k}{k!} + \cdots\right].$$

This latter series converges to

$$\frac{n-1}{n} + \frac{1}{n}e^{nM},$$

and the series S_{ij} converges absolutely. As a result, since for every finite-order matrix A a majorant can be found such that $|a_{ij}| \leqslant M$, the matrix $\exp(A)$ is completely defined for every matrix A of order n.

3. Let us assume that the matrices A and B commute. From that we deduce

$$(A+B)^2 = A^2 + 2AB + B^2,$$

and by induction

$$(A+B)^m = \sum_{k=0}^{m} C_m^k A^k B^{m-k}.$$

On the other hand, for the matrix elements

$$(e^A E^B)_{ij} = \sum_{k=1}^{n} (e^A)_{ik} (e^B)_{kj}.$$

The two terms of the right-hand side are represented by series of the form S_{ij}, each of which is absolutely convergent. Thus we can carry out the multiplication of these two series. If we carry out the product of the matrices straightforwardly we obtain

$$\begin{aligned} e^A e^B &= \left(\sum_{k=0}^{\infty} \frac{A^k}{k!}\right)\left(\sum_{l=0}^{\infty} \frac{B^l}{l!}\right) \\ &= \sum_{m=0}^{\infty} \left(\sum_{k=0}^{m} C_m^k A^k B^{m-k}\right) \frac{1}{m!}. \end{aligned}$$

It is seen that the term $(A+B)^m$ calculated before appears, whence

$$\begin{aligned} e^A e^B &= \sum_{m=0}^{\infty} \frac{(A+B)^m}{m!} \\ &= e^{A+B}. \end{aligned}$$

EXERCISE **2.4**

A two-dimensional representation of the group of plane rotations $\boldsymbol{SO}(2)$ is given by the matrices

$$M(G_\alpha) = \begin{bmatrix} \cos\alpha & -\sin\alpha \\ \sin\alpha & \cos\alpha \end{bmatrix}.$$

1. Calculate the infinitesimal matrix $X^{(\alpha)}$ of the given representation.
2. By a direct calculation verify the following expression for the matrices $M(R_\alpha)$ of this representation

$$M(R_\alpha) = \exp(\alpha X^{(\alpha)}).$$

SOLUTION **2.4**

1. Differentiation of the matrix elements of $M(R_\alpha)$ at $\alpha = 0$ gives the infinitesimal matrix, that is

$$X^{(\alpha)} = \begin{bmatrix} 0 & -1 \\ 1 & 0 \end{bmatrix}.$$

2. The matrix $\exp(\alpha X^{(\alpha)})$ is defined by the series

$$\exp(\alpha X^{(\alpha)}) = \mathbb{1} + \frac{\alpha X^{(\alpha)}}{1!} + \frac{\alpha^2 (X^{(\alpha)})^2}{2!} + \cdots + \frac{\alpha^k (X^{(\alpha)})^k}{k!} + \cdots .$$

The successive powers of the matrix $X^{(\alpha)}$ gives us

$$\begin{array}{lll} (X^{(\alpha)})^2 = -\mathbb{1}, & (X^{(\alpha)})^4 = \mathbb{1}, & (X^{(\alpha)})^{2n} = (-1)^n \mathbb{1}, \\ (X^{(\alpha)})^3 = -X^{(\alpha)}, & (X^{(\alpha)})^5 = X^{(\alpha)}, & (X^{(\alpha)})^{2n+1} = (-1)^n X^{(\alpha)}. \end{array}$$

The matrix series $\exp(\alpha X^{(\alpha)})$ can be split into two series:

— One contains the even powers of α, say α^{2n}, which are multiplied by $(-1)^n \mathbb{1}$. Its sum is equal to $\mathbb{1} \cos\alpha$.

— The other consists of the odd powers of α, say α^{2n+1}, which are multiplied by $(-1)^n X^{(\alpha)}$. Its sum is equal to $X^{(\alpha)} \sin\alpha$.

Finally we obtain

$$\begin{aligned} \exp(\alpha X^{(\alpha)}) &= \mathbb{1} \cos\alpha + X^{(\alpha)} \sin\alpha \\ &= \begin{bmatrix} \cos\alpha & 0 \\ 0 & \cos\alpha \end{bmatrix} + \begin{bmatrix} 0 & -\sin\alpha \\ \sin\alpha & 0 \end{bmatrix} \\ &= \begin{bmatrix} \cos\alpha & -\sin\alpha \\ \sin\alpha & \cos\alpha \end{bmatrix} \\ &= M(R_\alpha). \end{aligned}$$

EXERCISE **2.5**

The matrices $M(R_\alpha)$ of the representation of the group $\boldsymbol{SO}(2)$ are related by the relation (2.1.47)

$$M(R_\alpha) X^{(\alpha)} = \frac{\mathrm{d}M(R_\alpha)}{\mathrm{d}\alpha} \qquad \text{with} \quad X^{(\alpha)} = \begin{bmatrix} 0 & -1 \\ 1 & 0 \end{bmatrix}.$$

Calculate the matrices $M(R_\alpha)$ by solving the differential system which the elements M_{ij} of these matrices satisfy.

SOLUTION **2.5**

The differential equation for the matrix $M(R_\alpha)$ may be written in the expanded form

$$\begin{bmatrix} M_{11} & M_{12} \\ M_{21} & M_{22} \end{bmatrix} \begin{bmatrix} 0 & -1 \\ 1 & 0 \end{bmatrix} = \begin{bmatrix} M_{12} & -M_{11} \\ M_{22} & -M_{21} \end{bmatrix} = \begin{bmatrix} \dfrac{\mathrm{d}M_{11}}{\mathrm{d}\alpha} & \dfrac{\mathrm{d}M_{12}}{\mathrm{d}\alpha} \\ \dfrac{\mathrm{d}M_{21}}{\mathrm{d}\alpha} & \dfrac{\mathrm{d}M_{22}}{\mathrm{d}\alpha} \end{bmatrix},$$

whence the differential system

$$\frac{\mathrm{d}M_{11}}{\mathrm{d}\alpha} = M_{12}, \qquad \frac{\mathrm{d}M_{12}}{\mathrm{d}\alpha} = -M_{11}, \qquad \frac{\mathrm{d}M_{21}}{\mathrm{d}\alpha} = M_{22}, \qquad \frac{\mathrm{d}M_{22}}{\mathrm{d}\alpha} = -M_{21}.$$

With $M[R(0)] = \mathbb{1}$ the initial conditions may be written

$$M_{11}(0) = M_{22}(0) = 1, \qquad M_{12}(0) = M_{21}(0) = 0.$$

The differential system satisfying these initial conditions has as its solution

$$M_{11} = M_{22} = \cos\alpha, \qquad M_{12} = -\sin\alpha, \qquad M_{21} = \sin\alpha.$$

We obtain the classical matrix of the group of plane rotations $\boldsymbol{SO}(2)$.

EXERCISE **2.6**

1. By expanding the expression $\exp[-i\frac{1}{2}\alpha\sigma_z]$, where σ_z is the Pauli matrix given by (1.3.12), prove that a second-order matrix representing an element of the group $\boldsymbol{SU}(2)$ is obtained.

2. Calculate the expression for the spinor $\eta = (\psi_1, \psi_2)$ transformed by the matrix $\exp[-i\frac{1}{2}\alpha\sigma_z]$. What is the expression for the transform for $\alpha = 2\pi$? For which angle of rotation do we obtain an identity spinor?

SOLUTION **2.6**

1. The matrix $\exp[-i\frac{1}{2}\alpha\sigma_z]$ has as its expansion in a series

$$\exp[-i\tfrac{1}{2}\alpha\sigma_z] = \mathbb{1} + \sum_{n=1}^{\infty}(-i\tfrac{1}{2}\alpha)^n \frac{1}{n!} \begin{bmatrix} 1 & 0 \\ 0 & -1 \end{bmatrix}^n.$$

The successive powers of σ_z have as their expressions

$$(\sigma_z)^2 = \mathbb{1}, \qquad (\sigma_z)^{2n} = \mathbb{1}, \qquad (\sigma_z)^{2n+1} = \sigma_z.$$

Consequently, the previous series can be split into two series, that is

$$\begin{aligned}\exp[-i\tfrac{1}{2}\alpha\sigma_z] &= \cos\tfrac{1}{2}\alpha\mathbb{1} - i\sin\tfrac{1}{2}\alpha\sigma_z \\ &= \begin{bmatrix} e^{-i\alpha/2} & 0 \\ 0 & e^{i\alpha/2} \end{bmatrix}.\end{aligned}$$

A matrix is obtained representing the elements of the group $\boldsymbol{SU}(2)$ which correspond to

$$a = \exp(-i\tfrac{1}{2}\alpha) \qquad \text{and} \qquad b = 0.$$

2. A spinor transformed by this matrix becomes

$$\begin{aligned}\eta' &= \begin{bmatrix} e^{-i\alpha/2} & 0 \\ 0 & e^{i\alpha/2} \end{bmatrix}\begin{bmatrix} \psi_1 \\ \psi_2 \end{bmatrix} \\ &= \begin{bmatrix} e^{-i\alpha/2}\psi_1 \\ e^{i\alpha/2}\psi_2 \end{bmatrix}.\end{aligned}$$

In particular, for a rotation $\alpha = 2\pi$ we have $e^{\pm i\pi} = -1$, whence

$$\eta' = (-\psi_1, -\psi_2) = -\eta.$$

A rotation through an angle of 2π about the axis Oz transforms a spinor into its opposite. It is a rotation through an angle of 4π which once again gives a spinor identical to itself.

3

Spinor Representation of $SO(3)$

3.1 The Rotation Group $\boldsymbol{SO}(3)$

3.1.1 Rotations About a Point

The rotations of physical space only form a group in the precise case that we restrict ourselves to the study of the group of rotations about every axis passing through a given point of the physical space. This point will be chosen as the coordinates' origin. Furthermore, we shall restrict ourselves to the group of proper orthogonal transformations, that is to say, those of which the determinant of the rotation matrices is equal to $+1$. This group is denoted $\boldsymbol{SO}(3)$.

Rotation Matrices

Each element of the group $\boldsymbol{SO}(3)$ can be characterized by three continuous parameters. In fact, an arbitrary rotation can be decomposed into three successive rotations through angles α, β, γ respectively about the axes Ox, Oy, Oz of a Cartesian system. We consider an 'active' rotation, that is to say, a rotation under which a given point is displaced in a fixed frame of reference. Rotations called 'passive' are rotations of the frame of reference. Consequently we shall consider solely the active rotations. A rotation through an angle α about Ox corresponds to the matrix

$$M[R_x(\alpha)] = \begin{bmatrix} 1 & 0 & 0 \\ 0 & \cos\alpha & -\sin\alpha \\ 0 & \sin\alpha & \cos\alpha \end{bmatrix}. \tag{3.1.1}$$

Rotations through an angle β about Oy and through an angle γ about Oz correspond, respectively, to

$$\begin{aligned} M[R_y(\beta)] &= \begin{bmatrix} \cos\beta & 0 & \sin\beta \\ 0 & 1 & 0 \\ -\sin\beta & 0 & \cos\beta \end{bmatrix}, \\ M[R_z(\gamma)] &= \begin{bmatrix} \cos\gamma & -\sin\gamma & 0 \\ \sin\gamma & \cos\gamma & 0 \\ 0 & 0 & 1 \end{bmatrix}. \end{aligned} \tag{3.1.2}$$

These matrices form part of the group of matrices of order 3 denotes $\boldsymbol{SO}(3)$. An arbitrary rotation can thus be represented by the matrix product

$$M[R(\alpha,\beta,\gamma)] = M[R_x(\alpha)]M[R_y(\beta)]M[R_z(\gamma)]. \tag{3.1.3}$$

3.1.2 The Infinitesimal Matrices of the Group

Let us consider the matrix (3.1.1) denoted $M[R_x(\alpha)]$ and describing rotation about the axis Ox. The corresponding infinitesimal rotation is

$$\begin{aligned} L_x &= \left[\frac{\mathrm{d}M[R_x(\alpha)]}{\mathrm{d}\alpha}\right]_{\alpha=0} \\ &= \begin{bmatrix} 0 & 0 & 0 \\ 0 & 0 & -1 \\ 0 & 1 & 0 \end{bmatrix}. \end{aligned} \tag{3.1.4}$$

The same infinitesimal matrices for rotations about the axes Oy, Oz are obtained from the matrices (3.1.2), that is,

$$L_y = \begin{bmatrix} 0 & 0 & 1 \\ 0 & 0 & 0 \\ -1 & 0 & 0 \end{bmatrix}, \qquad L_z = \begin{bmatrix} 0 & -1 & 0 \\ 1 & 0 & 0 \\ 0 & 0 & 0 \end{bmatrix}. \tag{3.1.5}$$

Properties of the Infinitesimal Matrices

Let us remark that the antisymmetry symbol (the Kronecker symbol) ε_{ijk} can be used for representing the elements of the infinitesimal matrices. Let us write L_1, L_2, L_3, respectively, for the infinitesimal matrices L_x, L_y, L_z. The values of these elements can be put into the following form

$$(L_k)_{ij} = -\varepsilon_{ijk} \qquad i, j, k = 1, 2, 3. \tag{3.1.6}$$

It is easily verified that these matrices satisfy the commutation relations

$$L_x L_y - L_y L_x = L_z, \quad L_y L_z - L_z L_y = L_x, \quad L_z L_x - L_x L_z = L_y. \tag{3.1.7}$$

These relations determine the values of the structure constants defined by (2.1.34). Formula (2.1.48) gives us for the expression for the different rotation matrices,

$$\begin{aligned} M[R_x(\alpha)] &= \exp(\alpha L_x), \\ M[R_y(\beta)] &= \exp(\beta L_y), \\ M[R_z(\gamma)] &= \exp(\gamma L_z), \end{aligned} \tag{3.1.8}$$

The matrix corresponding to an arbitrary rotation $R(\alpha, \beta, \gamma)$ is then written as the product of the matrices (3.1.8), that is,

$$M[R(\alpha, \beta, \gamma)] = \exp(\alpha L_x) \exp(\beta L_y) \exp(\gamma L_z). \tag{3.1.9}$$

The infinitesimal matrices L_x, L_y, L_z do not commute with each other, and the classical relation $\exp(A)\exp(B) = \exp(A + B)$, where A and B are square matrices of the same order, cannot a priori be used.

Formula (2.1.37) for the restricted expansion of the matrices of a representation in the neighborhood of the unit matrix I may be written for the infinitesimal rotations $\mathrm{d}\alpha, \mathrm{d}\beta, \mathrm{d}\gamma$ as

$$M[R(\mathrm{d}\alpha, \mathrm{d}\beta, \mathrm{d}\gamma)] = I + \mathrm{d}\alpha L_x + \mathrm{d}\beta L_y + \mathrm{d}\gamma L_z. \tag{3.1.10}$$

The Infinitesimal Variation of a Vector

Let $\boldsymbol{r}$ be a vector defining the position of a point M of the coordinate space x, y, z. Under a rotation $R(\mathrm{d}\alpha, \mathrm{d}\beta, \mathrm{d}\gamma)$ through an infinitesimal angle about a given axis this vector is transformed into a vector $\boldsymbol{r}'$ such that

$$\boldsymbol{r}' = \boldsymbol{r} + \mathrm{d}\boldsymbol{r} = \Gamma[R(\mathrm{d}\alpha, \mathrm{d}\beta, \mathrm{d}\gamma)]\boldsymbol{r}. \tag{3.1.11}$$

Let us write $[\boldsymbol{r}]$, $[\boldsymbol{r}']$, and $[\mathrm{d}\boldsymbol{r}]$ for the column matrices which have as elements the components of the respective vectors $\boldsymbol{r}, \boldsymbol{r}'$, and $\mathrm{d}\boldsymbol{r}$. For a rotation close to the identity operation the relation (3.1.11) may be written in the matrix form

$$\begin{aligned}[\boldsymbol{r}'] &= [\boldsymbol{r}] + [\mathrm{d}\boldsymbol{r}] \\ &= (I + \mathrm{d}\alpha L_x + \mathrm{d}\beta L_y + \mathrm{d}\gamma L_z)[\boldsymbol{r}],\end{aligned} \tag{3.1.12}$$

whence the expression for the matrix of the elementary variations of the components of the vector

$$[\mathrm{d}\boldsymbol{r}] = [\mathrm{d}\alpha L_x + \mathrm{d}\beta L_y + \mathrm{d}\gamma L_z][\boldsymbol{r}]. \tag{3.1.13}$$

Let us consider an infinitesimal rotation through an angle $\mathrm{d}\alpha$ about the axis Ox. We obtain

$$\begin{aligned}[\mathrm{d}_\alpha \boldsymbol{r}] &= \mathrm{d}\alpha L_x[\boldsymbol{r}] \\ &= \mathrm{d}\alpha \begin{bmatrix} 0 & 0 & 0 \\ 0 & 0 & -1 \\ 0 & 1 & 0 \end{bmatrix} \begin{bmatrix} x \\ y \\ z \end{bmatrix} \\ &= \begin{bmatrix} 0 \\ -\mathrm{d}\alpha z \\ \mathrm{d}\alpha y \end{bmatrix}.\end{aligned} \tag{3.1.14}$$

Rotations $\mathrm{d}\beta$ and $\mathrm{d}\gamma$ respectively about the axes Oy and Oz give us similarly

$$\begin{aligned}[\mathrm{d}_\beta \boldsymbol{r}] = \mathrm{d}\beta L_y[\boldsymbol{r}] &= \begin{bmatrix} \mathrm{d}\beta z \\ 0 \\ -\mathrm{d}\beta x \end{bmatrix}, \\ [\mathrm{d}_\gamma \boldsymbol{r}] = \mathrm{d}\gamma L_z[\boldsymbol{r}] &= \begin{bmatrix} -\mathrm{d}\gamma y \\ \mathrm{d}\gamma x \\ 0 \end{bmatrix}.\end{aligned} \tag{3.1.15}$$

3.1.3 Rotations About a Given Axis

Let us consider a rotation through an angle θ about a given axis Δ with direction cosines (a, b, c) in an orthonormal reference frame. The rotation matrix may be written as a function of these parameters

$$M[R(\Delta,\theta)] = \cos\theta \begin{bmatrix} 1 & 0 & 0 \\ 0 & 1 & 0 \\ 0 & 0 & 1 \end{bmatrix} + (1-\cos\theta) \begin{bmatrix} a^2 & ab & ac \\ ba & b^2 & bc \\ ca & cb & c^2 \end{bmatrix}$$

$$+\sin\theta \begin{bmatrix} 0 & -c & b \\ c & 0 & -a \\ -b & a & 0 \end{bmatrix}. \tag{3.1.16}$$

The infinitesimal matrix $X[R(\Delta, \mathrm{d}\theta)]$ expressed as a function of the given parameters is immediately deduced from expression (3.1.16) by differentiating the matrix elements with respect to θ and taking their values at $\theta = 0$, whence

$$X[r(\Delta, \mathrm{d}\theta)] = \begin{bmatrix} 0 & -c & b \\ c & 0 & -a \\ -b & a & 0 \end{bmatrix}. \tag{3.1.17}$$

The matrix corresponding to the infinitesimal rotation $\mathrm{d}\theta$ about the given axis Δ may then be written

$$M[R(\Delta, \mathrm{d}\theta)] = I + \mathrm{d}\theta X[r(\Delta, \mathrm{d}\theta)]. \tag{3.1.18}$$

Under such a rotation a vector $\boldsymbol{r}$ is transformed into a vector $\boldsymbol{r}'$ given by

$$\begin{aligned} \boldsymbol{r}' &= (I + \mathrm{d}\theta X[R(\Delta, \mathrm{d}\theta)])\boldsymbol{r} \\ &= \boldsymbol{r} + \mathrm{d}\boldsymbol{r}. \end{aligned} \tag{3.1.19}$$

Let us write $\boldsymbol{u}$ for the unitary vector of the rotation axis with components (a, b, c) in an orthonormal frame. The calculation of the components of $\mathrm{d}\boldsymbol{r}$ with the aid of the matrix (3.1.17) shows that we have

$$\mathrm{d}\boldsymbol{r} = (\boldsymbol{u} \times \boldsymbol{r})\mathrm{d}\theta. \tag{3.1.20}$$

This latter expression can be recovered classically with the help of a geometric argument. Let us consider a unitary vector $\boldsymbol{u}$ along the rotation axis Δ (Figure 3.1). Let $\boldsymbol{OM}$ be a vector causing a rotation through an infinitesimal angle $\mathrm{d}\theta$ about Δ. Let us write α for the angle between the vectors $\boldsymbol{u}$ and $\boldsymbol{OM}$. In the course of this infinitesimal displacement the vector is transformed into $\boldsymbol{OM}'$. The variation $\mathrm{d}\boldsymbol{OM}$ of the vector is tangential to the circle of radius $\|\boldsymbol{OM}\| \sin\alpha$, and the vector product $\boldsymbol{u} \times \boldsymbol{OM}$ is perpendicular to the plane defined by $\boldsymbol{u}$ and $\boldsymbol{OM}$. By identifying the length of the arc of the curve with the norm of $\mathrm{d}\boldsymbol{OM}$ we obtain formula (3.1.20)

$$\boldsymbol{OM}' = \mathrm{d}\boldsymbol{OM} = \mathrm{d}\theta(\boldsymbol{u} \times \boldsymbol{OM}). \tag{3.1.21}$$

Equivalent Rotations About the Axes of a Reference Frame

Let us calculate the angles of rotation $\mathrm{d}\alpha, \mathrm{d}\beta, \mathrm{d}\gamma$ respectively about the axes Ox, Oy, Oz corresponding to a rotation $\mathrm{d}\theta$ about a given axis Δ passing

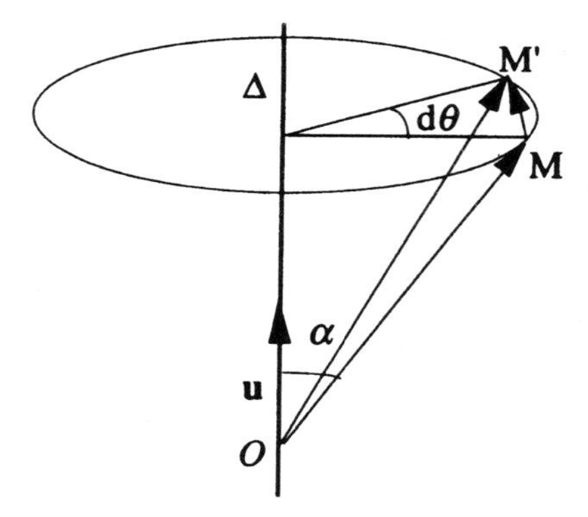

FIGURE 3.1. The infinitesimal displacement as a function of the angle of rotation.

through the origin of the reference system. The vector $\mathrm{d}\boldsymbol{r}$ given by (3.1.20) must then be equal to that calculated by (3.1.13), that is to say, on writing the vectors in column matrix form

$$[\mathrm{d}\boldsymbol{r}] = \mathrm{d}\theta[\boldsymbol{u} \times \boldsymbol{r}] = [\mathrm{d}\alpha L_x + \mathrm{d}\beta L_y + \mathrm{d}\gamma L_z][\boldsymbol{r}]. \tag{3.1.22}$$

Let us identify the components of the vector $\mathrm{d}\boldsymbol{r}$ calculated from the two previous expressions. Using (3.1.14) and (3.1.15) we obtain

$$a\mathrm{d}\theta = \mathrm{d}\alpha, \qquad b\mathrm{d}\theta = \mathrm{d}\beta, \qquad c\mathrm{d}\theta = \mathrm{d}\gamma. \tag{3.1.23}$$

Let us write $\boldsymbol{L}$ for the vector which has as its components the infinitesimal matrices L_x, L_y, L_z. In accordance with (3.1.23) we obtain

$$\begin{aligned} [\mathrm{d}\boldsymbol{r}] &= \mathrm{d}\theta(aL_x + bL_y + cL_z)[\boldsymbol{r}] \\ &= \mathrm{d}\theta[\boldsymbol{u}\cdot\boldsymbol{L}][\boldsymbol{r}]. \end{aligned} \tag{3.1.24}$$

Comparing this with the expression for $\mathrm{d}\boldsymbol{r}$ given by (3.1.19) we obtain as the expression for the infinitesimal matrix

$$X[R(\Delta, \mathrm{d}\theta)] = aL_x + bL_y + cL_z = \boldsymbol{u}\cdot\boldsymbol{L}. \tag{3.1.25}$$

This latter expression can also be obtained from the matrix (3.1.17), which can be easily decomposed into the three infinitesimal matrices (3.1.4) and (3.1.5).

3.1.4 The Exponential Matrix of a Rotation About a Given Axis

Rotation About the Oz Axis

Let us return to a more direct proof of expressions (3.1.8) corresponding to rotations about the axes of a Cartesian frame of reference. Let us consider a rotation through an angle $(\gamma + \mathrm{d}\gamma)$ about the axis Oz. Let us write $R(Oz, \gamma + \mathrm{d}\gamma)$ for this rotation, of which the matrix is written on taking account of (2.1.29)

$$\begin{aligned} M[R(Oz, \gamma + \mathrm{d}\gamma)] &= M[R(Oz, \gamma)]M[R(Oz, \mathrm{d}\gamma)] \\ &= M[R(Oz, \gamma)][I + \mathrm{d}\gamma L_z]. \end{aligned} \tag{3.1.26}$$

The derivative of the matrix $M[R(Oz, \gamma)]$ is introduced in the form

$$\frac{M[R(Oz, \gamma + \mathrm{d}\gamma)] - M[R(Oz, \gamma)]}{\mathrm{d}\gamma} = M[R(O\gamma)]L_z. \tag{3.1.27}$$

This is a matrix differential equation which can be completed with the obvious initial condition

$$M[R(Oz, \gamma = 0)] = I. \tag{3.1.28}$$

Definition (2.1.49) of the exponential matrix shows that this matrix satisfies the differential equation (3.1.27), whence

$$M[R(Oz, \gamma)] = \exp[\gamma L_z]. \tag{3.1.29}$$

Rotation About an Arbitrary Axis

For the moment let us use expression (3.1.25) for the infinitesimal rotation matrix about an axis along the unitary vector $\boldsymbol{u}$. The proof is analogous to that used for a rotation about the axis Oz, replacing the infinitesimal matrix L_z with the matrix $[\boldsymbol{u}\cdot\boldsymbol{L}]$. Thus we obtain the differential equation

$$\frac{M[R(\Delta, \theta + \mathrm{d}\theta)] - M[R(\Delta, \theta)]}{\mathrm{d}\theta} = M[R(\Delta, \theta)][\boldsymbol{u}\cdot\boldsymbol{L}]. \tag{3.1.30}$$

As the matrix $\boldsymbol{u}\cdot\boldsymbol{L}$ is independent of the angle of rotation θ integration gives us

$$\begin{aligned} M[R(\Delta, \theta)] &= \exp(\theta \boldsymbol{u}\cdot\boldsymbol{L}) \\ &= \exp[\theta(aL_x + bL_y + cL_z)], \end{aligned} \tag{3.1.31}$$

where a, b, c are the components of the unitary vector $\boldsymbol{u}$ defining the axis of rotation.

3.2 Irreducible Representations of $\boldsymbol{SO}(3)$

3.2.1 *The Structure Equations*

In what follows we are going to calculate all the irreducible representations of $\boldsymbol{SO}(3)$. To do so, first of all we are going to prove that all the infinitesimal matrices A_x, A_y, A_z of the operators associated with the infinitesimal rotations of an arbitrary linear representation of **SO**(3) satisfy the same commutation relations (3.1.7) as the matrices L_x, L_y, L_z. We are thus going to prove a particular case of a consequence of Sophus Lie's theorem stated in the form of relation (2.1.26).

Let us consider two rotations, one a rotation through an angle α about an axis with direction cosines (a, b, c), and the other a rotation through an angle β about the axis (a', b', c'). Let us write $M(\alpha)$ and $M(\beta)$ for the matrices which represent the rotation operators which operate on the vectors of the Euclidean space E_3. Let us form the following matrix:

$$M = M(\alpha)M(\beta)M(-\alpha)M(-\beta). \tag{3.2.1}$$

This matrix is expandable in powers of α and β. It is reduced to the unit matrix $\mathbb{1}$ for $\alpha = \beta = 0$, and it is also reduced to the unit matrix for $\alpha = 0$ and $\beta = 0$. As a consequence all the terms of the expansion apart from the first must contain $\alpha\beta$ as a factor. The first-order term of the matrix $M - \mathbb{1}$ will therefore be of the form $\alpha\beta P$, where P is a matrix which represents an infinitesimal rotation. In order to obtain the matrix P it suffices to carry out the product (3.2.1) restricting each factor to its first term, that is,

$$\begin{aligned} M(\alpha) &= \mathbb{1} + \alpha P_1, & M(\beta) &= \mathbb{1} + \beta P_2, \\ M(-\alpha) &= \mathbb{1} - \alpha P_1, & M(-\beta) &= \mathbb{1} - \beta P_2. \end{aligned} \tag{3.2.2}$$

The product of these matrices gives us

$$M = \mathbb{1} + \alpha\beta(P_1P_2 - P_2P_1) + \cdots . \tag{3.2.3}$$

Thus we end up with the following proposition: if P_1 and P_2 are matrices respectively representing two infinitesimal operations operating on the vectors of the Euclidean space E_3, the matrix $(P_1P_2 - P_2P_3)$ also represents an infinitesimal rotation.

If we consider, for the moment, an arbitrary linear representation of the rotation group—in other words, if instead of having E_3 for the representation space we consider a representation space V_n—we can affirm, in the same way in which we obtain the matrix $(P_1P_2 - P_2P_1)$, that to the infinitesimal rotation $(P_1P_2 - P_2P_1)$ there corresponds the new representation matrix $(A_1A_2 - A_2A_1)$ by calling A_1 and A_2 the matrices which correspond to P_1 and P_2.

In particular, if A_x, A_y, A_z are the matrices which are associated with the infinitesimal rotations of a linear representation, then these three matrices satisfy the same relations (3.1.7) which are satisfied by the matrices L_x, L_y, L_z. In other words, we have

$$\begin{aligned} A_xA_y - A_yA_x &= A_z, \\ A_yA_z - A_zA_y &= A_x, \\ A_zA_x - A_zA_z &= A_y. \end{aligned} \tag{3.2.4}$$

The structure coefficients defined in Sophus Lie's theorem (2.1.26) are equal to unity or zero in this present case.

3.2.2 The Infinitesimal Matrices of the Representations of the Group $\boldsymbol{SO}(3)$

We have just seen that the infinitesimal matrices A_x, A_y, A_z of an arbitrary representation satisfy the structure relations of $\boldsymbol{SO}(3)$. These relations are going to allow us to determine explicitly the infinitesimal matrices of all the irreducible representations of the rotation group. Next, the matrices $M[R(\alpha, \beta, \gamma)]$ of the representations of the group will be able to be obtained from the general expression

$$M[R(\alpha,\beta,\gamma)] = \exp(\alpha A_x + \beta A_y + \gamma A_z). \tag{3.2.5}$$

In order to determine the matrices A_x, A_y, A_z explicitly we are going to use the following linear combinations of matrices the properties of which are easier to study

$$A_+ = iA_x - A_y, \qquad A_- = iA_x + A_y, \qquad A_3 = iA_z. \tag{3.2.6}$$

The commutation relations (3.2.4) give the following commutation relations

$$[A_+, A_-] = 2A_3, \qquad [A_-, A_3] = A_-, \qquad [A_3, A_+] = A_+, \tag{3.2.7}$$

as can be easily verified.

Anti-Hermiticity

As the group $\boldsymbol{SO}(3)$ is compact we prove that every representation of this group is equivalent to a unitary representation. We shall therefore assume that the matrices $M[R(\alpha, \beta, \gamma)]$ are those of a unitary representation, whence,

$$M^\dagger[R(\alpha,\beta,\gamma)]M[R(\alpha,\beta,\gamma)] = \mathbb{1}. \tag{3.2.8}$$

Let us carry out a limited expansion of this latter expression in the neighborhood of the identity matrix, and preserving only the terms of the first degree in α, β, γ. This yields

$$\begin{aligned}(\mathbb{1} + \alpha {A_x}^\dagger + \beta {A_y}^\dagger + \gamma {A_z}^\dagger)&(\mathbb{1} + \alpha A_x + \beta A_y + \gamma A_z) \\ &= \mathbb{1} + \alpha({A_x}^\dagger + A_x) + \beta({A_y}^\dagger + A_y) + \gamma({A_z}^\dagger + A_z) + \cdots \\ &= \mathbb{1},\end{aligned} \tag{3.2.9}$$

whence

$$ {A_x}^\dagger + A_x, \qquad {A_y}^\dagger - A_y, \qquad {A_z}^\dagger = -A_z. \tag{3.2.10}$$

The infinitesimal matrices of an arbitrary representation are anti-Hermitian. Let us write $\boldsymbol{x}\cdot\boldsymbol{y}$ for the product of the two vector $\boldsymbol{x}$ and $\boldsymbol{y}$. The properties (3.2.10) of anti-Hermiticity give us

$$A_+\boldsymbol{x}\cdot\boldsymbol{y} = \boldsymbol{x}\cdot A_-\boldsymbol{y}. \tag{3.2.11}$$

3.2.3 Eigenvectors and Eigenvalues of the Infinitesimal Matrices of the Representations

Let $\boldsymbol{x}_\lambda$ be an eigenvector of the matrix A_3 for the eigenvalue λ. This vector belongs to the representation space, because by hypothesis A_3 is a matrix of this representation. We shall also denote by $\boldsymbol{x}_\lambda$ the column matrix representing this vector, and we shall write

$$A_3\boldsymbol{x}_\lambda = \lambda\boldsymbol{x}_\lambda. \tag{3.2.12}$$

Let us show that $A_+\boldsymbol{x}_\lambda$ is an eigenvector of the matrix A_3 for the eigenvalue $(\lambda+1)$. In fact, according to the third relation of (3.2.7) we have

$$\begin{aligned} A_3(A_+\boldsymbol{x}_\lambda) &= (A_+A_3 + A_+)\boldsymbol{x}_\lambda \\ &= A_+(A_3\boldsymbol{x}_\lambda) + A_+\boldsymbol{x}_\lambda \\ &= (\lambda+1)A_+\boldsymbol{x}_\lambda. \end{aligned} \tag{3.2.13}$$

Using the second commutation relation of (3.2.7) we also obtain

$$A_3(A_-\boldsymbol{x}_\lambda) = (A_-A_3 - A_-)\boldsymbol{x}_\lambda = (\lambda-1)A_-\boldsymbol{x}_\lambda, \tag{3.2.14}$$

which shows that $A_-\boldsymbol{x}_\lambda$ is an eigenvector of the matrix A_3 for the eigenvalue $(\lambda-1)$. Since the matrix A_3 is Hermitian all the eigenvalues are real, and two eigenvectors for two distinct eigenvalues are orthogonal.

Eigenvalues of the Matrix A_3

Let us assume that the eigenvectors $\boldsymbol{x}_\lambda$ have been orthonormalized. Let us write as $\boldsymbol{x}_{\lambda+1}$ and $\boldsymbol{x}_{\lambda-1}$ for the eigenvectors of A_3 for the eigenvalues $\lambda+1$ and $\lambda-1$. Relations (3.2.13) and (3.2.14) allow us to write

$$A_+\boldsymbol{x}_\lambda = \beta_\lambda\boldsymbol{x}_{\lambda+1}, \qquad A_-\boldsymbol{x}_\lambda = \alpha_\lambda\boldsymbol{x}_{\lambda-1}, \tag{3.2.15}$$

where α_λ and β_λ are the numbers which we are going to calculate. Taking account of property (3.2.11) of the Hermitian product we obtain for the orthonormal eigenvectors of A_3

$$\begin{aligned} \langle A_+\boldsymbol{x}_\lambda, \boldsymbol{x}_{\lambda+1}\rangle &= \beta_\lambda\langle \boldsymbol{x}_{\lambda+1}, \boldsymbol{x}_{\lambda+1}\rangle = \beta_\lambda \\ &= \langle \boldsymbol{x}_\lambda, A_-\boldsymbol{x}_{\lambda+1}\rangle \\ &= \alpha_{\lambda+1}\langle \boldsymbol{x}_\lambda, \boldsymbol{x}_\lambda\rangle = \alpha_{\lambda+1}. \end{aligned} \tag{3.2.16}$$

This latter relation gives us

$$\beta_\lambda = \alpha_{\lambda+1}. \tag{3.2.17}$$

In order to calculate the coefficients let us look for a recurrence relation between the values of β_λ. The first commutation relation of (3.2.7) gives us

$$\begin{aligned} A_+\boldsymbol{x}_{\lambda-1} &= A_+\left(\frac{1}{\alpha_\lambda}A_-\boldsymbol{x}_\lambda\right) \\ &= \frac{1}{\alpha_\lambda}(A_-A_+ + 2A_3)\boldsymbol{x}_\lambda \\ &= \frac{1}{\alpha_\lambda}(\alpha_{\lambda+1}\beta_\lambda + 2\lambda)\boldsymbol{x}_\lambda \\ &= \beta_{\lambda-1}\boldsymbol{x}_\lambda. \end{aligned} \tag{3.2.18}$$

In (3.2.18) let us replace the coefficient α by β in accordance with (3.2.7). This yields

$$(\beta_\lambda)^2 + 2\lambda = (\beta_{\lambda-1})^2. \tag{3.2.19}$$

On the other hand, the application of A_+ to an eigenvector $\boldsymbol{x}_\lambda$ allows us to obtain the vector $\boldsymbol{x}_{\lambda+1}$ corresponding to the eigenvalue $\lambda+1$. This process cannot be continued indefinitely because for a matrix A_3 of a chosen representation the number of independent vectors of the representation space is bounded. Therefore if we write j for the largest eigenvalue of A_3, the only possibility for the vector $\boldsymbol{x}_j$ is

$$A_+\boldsymbol{x}_j = \boldsymbol{0} = \beta_j\boldsymbol{x}_j, \tag{3.2.20}$$

so $\beta_j = 0$. For $\lambda = j$ relation (3.2.19) can then be written

$$(\beta_{j-1})^2 = 2j, \tag{3.2.21}$$

whence, by using (3.2.19) and (3.2.21), the following relation obtained by induction

$$(\beta_\lambda)^2 = j(j+1) - \lambda(\lambda+1). \tag{3.2.22}$$

Relation (3.2.17) also gives us

$$(a_\lambda)^2 = j(j+1) - \lambda(\lambda-1). \tag{3.2.23}$$

The application of the matrices A_+ and A_- to the eigenvector $\boldsymbol{x}_\lambda$ may then be written

$$\begin{aligned} A_+\boldsymbol{x}_\lambda &= \sqrt{j(j+1) - \lambda(\lambda+1)}\,\boldsymbol{x}_{\lambda+1}, \\ A_-\boldsymbol{x}_\lambda &= \sqrt{j(j+1) - \lambda(\lambda-1)}\,\boldsymbol{x}_{\lambda-1}. \end{aligned} \tag{3.2.24}$$

For $\lambda = -j$ the second formula of (3.2.15) gives us $A_- \boldsymbol{x}_{-j} = \mathbf{0}$. Starting from the eigenvector $\boldsymbol{x}_j$ the matrix A_- gives by successive application the sequence of vectors $\boldsymbol{x}_{j-1}, \boldsymbol{x}_{j-2}$, etc. The last vector of this sequence is $\boldsymbol{x}_{-j}$, since it must satisfy the relation $A_- \boldsymbol{x}_{-j} = \mathbf{0}$ exactly. Since j is the largest eigenvalue of A_3 its smallest value is therefore equal to $-j$. The eigenvalues vary from unity when it moves from a vector $\boldsymbol{x}_{j-k}$ to $\boldsymbol{x}_{j-k-1}$. Consequently $2j$ is an integer number and j must be equal to an integer number or half an integer (i.e., a half-integer). The number of eigenvectors is equal to $(2j+1)$.

3.2.4 Irreducible Representations

The orthonormal eigenvectors of the matrix A_3 form a sequence $\boldsymbol{x}_j, \boldsymbol{x}_{j-1}, \ldots, \boldsymbol{x}_{-j}$ and constitute a basis of a $(2j+1)$-dimensional vector space. The latter can act as a basis of the representation for the groups $\boldsymbol{SO}(3)$. In fact, formulas (3.2.12) and (3.2.24) show that the matrices A_3, A_+ and A_- leave E_{2j+1} invariant, and the same holds for the matrices $M[(R(\alpha, \beta, \gamma)]$ which can be expressed as a function of the infinitesimal matrices.

Let us show that the canonical basis gives a representation which is irreducible. To do so it suffices to show that the space E_{2j+1} does not contain subspaces invariant under rotations with representation matrices $M[R(\alpha, \beta, \gamma)]$. Let us assume the contrary, that is to say, let us assume that there exists a space E_k which is a vector subspace of E_{2j+1} invariant under the transformations $M[R(\alpha, \beta, \gamma)]$ and thus also under the infinitesimal matrices A_3, A_+, A_-. Let $\boldsymbol{y}$ be a vector of E_k for the largest eigenvalue of A_3. Since $\boldsymbol{y}$ is a vector of E_{2j+1} it can be written

$$\boldsymbol{y} = \sum_{k=-j}^{j} a_k \boldsymbol{x}_k. \tag{3.2.25}$$

By the hypothesis we must have $A_+ \boldsymbol{y} = \mathbf{0}$, whence

$$\begin{aligned} A_+ \boldsymbol{y} &= \sum_{k=-j}^{h} a_k A_+ \boldsymbol{x}_k \\ &= \sum_{k=-j}^{j} a_k \beta_k \boldsymbol{x}_{k+1} = \mathbf{0}. \end{aligned} \tag{3.2.26}$$

As the vectors $\boldsymbol{x}_k$ are linearly independent we must have

$$a_{-j} = a_{-j+1} = \cdots = a_{j-1} = 0. \tag{3.2.27}$$

Equation (3.2.25) thus gives us $\boldsymbol{y} = a_j \boldsymbol{x}_j$, which shows that $\boldsymbol{x}_j$ is an element of E_k. But since applying A_- to $\boldsymbol{x}_j$ gives the sequence of eigenvectors $\boldsymbol{x}_k$, the latter belong to E_k, and as a result E_k coincides with E_{2j+1}.

This proves the irreducibility of the representation realized in the canonical basis.

An irreducible representation of $\boldsymbol{SO}(3)$ is thus specified by the specification of j. This number is called the WEIGHT *of the irreducible representation.* We shall write $\Gamma^{(j)}$ for the irreducible representation of weight j, and the dimension of which is $2j+1$.

To a given dimension $n = 2j+1$ there corresponds at most one linear irreducible linear representation. As we have proved the existence of an irreducible representation for all the value of n, it follows that there exist no other irreducible representations.

The Matrix A^2

Let us denote by A^2 the matrix defined by

$$A^2 = {A_x}^2 + {A_y}^2 + {A_z}^2, \tag{3.2.28}$$

and let us show that the vectors of the canonical basis are eigenvectors of A^2 for the eigenvalue $-j(j+1)$. To do so let us write A^2 as a function of the matrices A_+, A_-, and A_3. We obtain

$$A^2 = -A_+A_- + A_3 - {A_3}^2. \tag{3.2.29}$$

The various relations obtained previously give us

$$\begin{aligned} A^2\boldsymbol{x}_\lambda &= (-A_+A_- + A_3 - {A_3}^2)\boldsymbol{x}_\lambda \\ &= (-\beta_{\lambda-1^2} + \lambda - \lambda^2)\boldsymbol{x}_\lambda, \end{aligned} \tag{3.2.30}$$

whence, on taking (3.2.22) into account,

$$A^2\boldsymbol{x}_\lambda = -j(j+1)\boldsymbol{x}_\lambda, \qquad \lambda = -j, \ldots, j. \tag{3.2.31}$$

3.2.5 The Infinitesimal Matrices of an Irreducible Representation in the Canonical Basis

Definition (3.2.10) of the matrices A_+, A_-, and A_3 allow us to calculate the matrices A_x, A_y, and A_3 as functions of the preceding ones. Relations (3.2.12) and (3.2.15) then give us

$$\begin{aligned} A_x\boldsymbol{x}_\lambda &= -\tfrac{1}{2}i(A_+ + A_-)\boldsymbol{x}_\lambda \\ &= -\tfrac{1}{2}i(\alpha_{\lambda+1}\boldsymbol{x}_{\lambda+1} + \alpha_\lambda\boldsymbol{x}_{\lambda-1}), \\ A_y\boldsymbol{x}_\lambda &= \tfrac{1}{2}(A_- - A_+)\boldsymbol{x}_\lambda \\ &= \tfrac{1}{2}(\alpha_\lambda\boldsymbol{x}_{\lambda-1} - \alpha_{\lambda+1}\boldsymbol{x}_{\lambda+1}), \\ A_x\boldsymbol{x}_\lambda &= -i\lambda\boldsymbol{x}_\lambda. \end{aligned} \tag{3.2.32}$$

We obtain the following Hermitian products for the orthonormal vectors

$$\begin{aligned}(A_x)_{kl} &= \langle \boldsymbol{x}_k, A_x \boldsymbol{x}_l \rangle \\ &= -\tfrac{1}{2} i (\alpha_{l+1} \delta_{k.l+1} + \alpha_l \delta_{k,l-1}), \\ (A_y)_{kl} &= \langle \boldsymbol{x}_k, A_y \boldsymbol{x}_l \rangle \\ &= \tfrac{1}{2} (\alpha_\lambda \delta_{k,l-1} - \alpha_{l+1} \delta_{k,l+1}), \\ (A_z)_{kl} &= \langle \boldsymbol{x}_k, A_z \boldsymbol{x}_l \rangle \\ &= -il\delta_{kl},\end{aligned} \tag{3.2.33}$$

with $k, l = -j, \ldots, j$. Let us classify these quantities in a matrix form by taking as the order of the vectors $\boldsymbol{x}_k$ the values $k = j, \ldots, -j$. We obtain the following matrices of order $2j + 1$:

The Infinitesimal Matrices of an Irreducible Representation in the Canonical Basis

$$A_x{}^{(j)} = \begin{bmatrix} 0 & -i\frac{1}{2}\alpha_j & \ldots & 0 & 0 \\ -i\frac{1}{2}\alpha_j & 0 & \ldots & 0 & 0 \\ \vdots & \vdots & \vdots & \vdots & \vdots \\ 0 & 0 & \ldots & 0 & -i\frac{1}{2}\alpha_{-j+1} \\ 0 & 0 & \ldots & -i\frac{1}{2}\alpha_{-j+1} & 0 \end{bmatrix},$$

$$A_y^{(j)} = \begin{bmatrix} 0 & -\frac{1}{2}\alpha_j & \ldots & 0 & 0 \\ \frac{1}{2}\alpha_j & 0 & \ldots & 0 & 0 \\ \vdots & \vdots & \vdots & \vdots & \vdots \\ 0 & 0 & \ldots & 0 & -\frac{1}{2}\alpha_{-j+1} \\ 0 & 0 & \ldots & \frac{1}{2}\alpha_{-j+1} & 0 \end{bmatrix}, \tag{3.2.34}$$

$$A_x{}^{(j)} = \begin{bmatrix} -ij & 0 & \ldots & 0 \\ 0 & i(-j+1) & \ldots & 0 \\ \vdots & \vdots & \vdots & \vdots \\ 0 & 0 & \ldots & ij \end{bmatrix}.$$

REMARK **3.1.** In quantum mechanics we calculate the matrices of the operators representing the components of the 'kinetic energy' analogues of the rotation operators to within a multiplicative factor. As a result, in order to obtain the irreducible representations, classically denoted in quantum mechanics as J_x, J_y, J_z, it is necessary to multiply the matrices A_x, A_y, A_z given above by $i\hbar$. ■

3.2.6 The Characters of the Rotation Matrices of a Representation

The matrices $M^{(j)}[R(\alpha,\beta,\gamma)]$ of the representation $\Gamma^{(j)}$ can be obtained by (3.2.5). Let us calculate the matrix $M^{(j)}[R(0,0,\varphi)]$ in the canonical basis. This yields

$$M^{(j)}[R(0,0,\varphi)] = \exp(\varphi A_z). \tag{3.2.35}$$

The latter matrix had as its definition

$$\exp(\varphi A_z) = I + \frac{\varphi}{1!}A_z + \frac{\varphi^2}{2!}A_z^2 + \frac{\varphi^3}{3!}A_z^3 + \cdots . \tag{3.2.36}$$

Since A_z is a diagonal matrix, by (3.2.25) all the powers of this matrix are diagonal and the matrix elements are obtained in the form of a series, which gives us after summation

$$M^{(j)}[R(0,0,\varphi)] = \begin{bmatrix} e^{-ij\varphi} & 0 & \dots & 0 \\ 0 & e^{i(-j+1)\varphi} & \dots & 0 \\ \vdots & \vdots & \vdots & \vdots \\ 0 & 0 & \dots & e^{ij\varphi} \end{bmatrix}. \tag{3.2.37}$$

The rotation $R(0,0,\varphi)$ corresponds to a rotation about the axis Oz and constitutes a representative of the class of rotations through an angle $|\varphi|$ about an arbitrary axis passing through the origin. The character of this class is obtain by forming the sum of the diagonal elements of the matrix (3.2.37), that is,

$$X^{(j)}(\varphi) = \sum_{k=-j}^{j} e^{ik\varphi} = \frac{\sin[j+\frac{1}{2}]\varphi}{\sin\frac{1}{2}\varphi}. \tag{3.2.38}$$

3.3 Spherical Harmonics

3.3.1 The Infinitesimal Operators in Spherical Coordinates

In the course of the previous Chapter 2 we have considered as $\boldsymbol{SO}(3)$ an arbitrary representation space E_{2j+1} with the eigenvectors $\boldsymbol{x}_\lambda$ as its canonical basis. Let us seek to realize E_{2j+1} as a subspace of the vector space of differentiable functions $f(\theta, \varphi)$ defined on the surface of a sphere of unit radius, θ and φ being the spherical coordinates.

We have seen that in quantum mechanics the operators of a transformation G are defined by considering their action on the wave functions, and which leads to the expression (2.1.14). For the moment let us consider the functions $f(\theta, \varphi)$ as wave functions. In the course of a rotation $R(\alpha, \beta, \gamma)$ the functions $f(\theta, \varphi)$ are transformed by the rotation operator $\Gamma(R)$ into the form

$$[\Gamma(R)f](\theta, \varphi) = f[R^{-1}(\theta, \varphi)]. \tag{3.3.1}$$

The operators $\Gamma(R)$ form a representation of the group $\boldsymbol{SO}(3)$ the matrices of which we have studied in the course of the previous chapter. We are going to calculate the infinitesimal operators L_x, L_y, L_z of this representation in spherical coordinates. For a rotation through an angle γ about the axis Oz relation (3.3.1) is written in the form of a limited expansion

$$\Gamma[R(0, 0, \alpha,)]f(\theta, \varphi) = f(\theta, \varphi - \gamma) - \gamma \frac{\partial R(\theta, \varphi)}{\partial \varphi} + \cdots .$$

The infinitesimal operator L_z thus has as its expression

$$L_z = -\frac{\partial}{\partial \varphi}. \tag{3.3.2}$$

For the moment let us consider a rotation through an angle α about the axis Ox. We then obtain the expansion

$$\begin{aligned} \Gamma[R(\alpha, 0, 0)]f(\theta, \varphi) &= f(\theta', \varphi') \\ &= f(\theta, \varphi) + \alpha \frac{\partial f}{\partial \theta} \left[\frac{\mathrm{d}\theta'}{\mathrm{d}\alpha}\right]_0 + \alpha \frac{\partial f}{\partial \varphi} \left[\frac{\mathrm{d}\varphi'}{\mathrm{d}\alpha}\right]_0 + \cdots . \end{aligned} \tag{3.3.3}$$

The infinitesimal operator has the form

$$L_x = \left[\frac{\mathrm{d}\theta'}{\mathrm{d}\alpha}\right]_0 \frac{\partial}{\partial \theta} + \left[\frac{\mathrm{d}\varphi'}{\mathrm{d}\alpha}\right]_0 \frac{\partial}{\partial \varphi}. \tag{3.3.4}$$

Let us consider the rotation of a vector $\boldsymbol{n}$ of unit length and with components

$$n_x = \sin\theta \cos\theta, \quad n_y = \sin\theta \sin\varphi, \quad n_z = \cos\theta. \tag{3.3.5}$$

After a rotation through α about the axis Ox the new components of $\boldsymbol{n}$ are

$$\begin{aligned} n_x{}' &= n_x = \sin\theta' \cos\varphi', \\ n_y{}' &= n_y \cos\alpha + n_z \sin\alpha = \sin\theta' \sin\varphi', \\ n_z{}' &= -n_y \sin\alpha + n_z \cos\alpha = \cos\theta'. \end{aligned} \tag{3.3.6}$$

We obtain

$$\left[\frac{\mathrm{d}n_x'}{\mathrm{d}\alpha}\right]_0 = 0, \qquad \left[\frac{\mathrm{d}n_y'}{\mathrm{d}\alpha}\right]_0 = n_z, \qquad \left[\frac{\mathrm{d}n_z'}{\mathrm{d}\alpha}\right]_{\alpha=0-} = -n_y. \tag{3.3.7}$$

Differentiating the expressions (3.3.6) written as functions of θ' and φ', then with $\theta' = \theta$ and $\varphi' = \varphi$ when $\alpha = 0$, this yields on equating with (3.3.7),

$$\begin{aligned}
\cos\theta\cos\varphi\left[\frac{\mathrm{d}\theta'}{\mathrm{d}\alpha}\right]_0 - \sin\theta\sin\varphi\left[\frac{\mathrm{d}\varphi'}{\mathrm{d}\alpha}\right]_0 &= 0,\\
\cos\theta\sin\varphi\left[\frac{\mathrm{d}\theta'}{\mathrm{d}\alpha}\right]_0 + \sin\theta\cos\varphi\left[\frac{\mathrm{d}\varphi'}{\mathrm{d}\alpha}\right]_0 &= \cos\theta,\\
\sin\theta\left[\frac{\mathrm{d}\theta'}{\mathrm{d}\alpha}\right]_0 &= \sin\theta\sin\varphi.
\end{aligned} \tag{3.3.8}$$

These relations give us

$$\left[\frac{\mathrm{d}\theta'}{\mathrm{d}\alpha}\right]_0 = \sin\varphi, \qquad \left[\frac{\mathrm{d}\varphi'}{\mathrm{d}\alpha}\right]_0 = \cot\theta\cos\varphi, \tag{3.3.9}$$

whence the infinitesimal operator L_x, using (3.3.4),

$$L_x = \sin\varphi\frac{\partial}{\partial\theta} + \cot\theta\cos\varphi\frac{\partial}{\partial\varphi}. \tag{3.3.10}$$

By carrying out a rotation through an angle β about the axis Oy and using the same method as above we obtain

$$L_y = -\cos\varphi\frac{\partial}{\partial\theta} + \cot\theta\sin\varphi\frac{\partial}{\partial\varphi}. \tag{3.3.11}$$

3.3.2 Spherical Harmonics

Let us seek amongst the functions $f(\theta, \varphi)$ those which will be able to serve as a basis for the irreducible representations of weight j of the group of spacial rotations. We have seen that the basis vectors $\boldsymbol{x}_\lambda$ of an irreducible representation must satisfy relation (3.2.31). For the operators corresponding to the infinitesimal matrices and a vector space formed by the functions $f(\theta, \varphi)$ this relation may be written

$$\begin{aligned}
L^2 f(\theta,\varphi) &= ({L_x}^2 + {L_y}^2 + {L_z}^2) f(\theta,\varphi)\\
&= -j(j+1) f(\theta,\varphi).
\end{aligned} \tag{3.3.12}$$

Moreover, if we want to calculate a canonical basis we have seen that the latter is formed of eigenvectors of the matrix A_z. In order to keep the classical notation of quantum mechanics we shall write this relation in the form

$$L_z f(\theta,\varphi) = -imf(\theta,\varphi), \qquad -j,\ldots,j. \tag{3.3.13}$$

Expressions (3.3.2), (3.3.10), and (3.3.11) of the infinitesimal operators allow us to write (3.3.12) and (3.3.13) explicitly. We obtain

$$\frac{1}{\sin\theta}\frac{\partial}{\partial\theta}\left[\sin\theta\frac{\partial f(\theta,\varphi)}{\partial\theta}\right] + \frac{1}{\sin^2\theta}\frac{\partial^2 f(\theta,\varphi)}{\partial\varphi^2} + j(j+1)f(\theta,\varphi) = 0, \tag{3.3.14}$$

$$\frac{\partial f(\theta,\varphi)}{\partial\varphi} = -imf(\theta,\varphi). \tag{3.3.15}$$

The calculation of the common eigenfunctions of the operators L^2 and L_z is achieved by looking for functions of the form

$$Y_{lm}(\theta,\varphi) = P_{lm}(\theta)\Phi_m(\varphi), \tag{3.3.16}$$

where the $Y_{lm}(\theta,\varphi)$ are called the *spherical harmonics*.

Equation (3.3.15) has

$$f(\theta,\varphi) = A(\theta)e^{-im\varphi}$$

as its general solution. We shall denote by $\Phi(\varphi)$ the part depending on φ, and which in a normalized form may be written

$$\Phi_m(\varphi) = \frac{1}{\sqrt{2\pi}}e^{im\varphi}. \tag{3.3.17}$$

In order that the function be unique it is necessary that m be an integer and that the condition (3.3.13) is recovered. Let us set $j = l$ in order to use the classical notation. Substituting expression (3.3.16) into equation (3.3.14), on taking (3.3.17) into account this yields

$$\frac{1}{\sin\theta}\frac{\mathrm{d}}{\mathrm{d}\theta}\left[\sin\theta\frac{\mathrm{d}P_{lm}(\theta)}{\mathrm{d}\theta}\right] - \frac{m^2}{\sin^2\theta}P_{lm}(\theta) + l(l+1)P_{lm}(\theta) = 0. \tag{3.3.18}$$

The latter equation only has solutions satisfying conditions of finiteness and uniqueness for the positive integers $l \geqslant |m|$. The corresponding solutions are called the *associated Legendre polynomials* $P_{lm}(\cos\theta)$, which have as their normalized expression

$$P_{lm}(\theta) = (-1)^m\left[\frac{(2l+1)(l-m)!}{2(l+m)!}\right]^{1/2}\frac{1}{2^l l!\sin^m\theta}\frac{\mathrm{d}^{l-m}\sin^{2l}\theta}{(\mathrm{d}\cos\theta)^{l-m}}, \tag{3.3.19}$$

with $m \geqslant 0$. For negative m the $P_{lm}(\theta)$ are defined by

$$P_{l,-|m|}(\theta) = (-1)^m P_{l|m|}(\theta), \tag{3.3.20}$$

where $P_{l|m|}(\theta)$ means that m is replaced by $|m|$ in formula (3.3.19).

We can show straightforwardly that the transformations of the spherical harmonics (3.3.16) under rotation give once again, for a given value of l, a linear combination of spherical harmonics corresponding to the various values of m running from $-l$ to l. From this it results that for a fixed l the Y_{lm} form a $(2l+1)$-dimensional vector space which is able to serve as a representation space for $\boldsymbol{SO}(3)$. Thus, as will be seen in the remainder, the representation obtained from the Y_{lm} is equivalent to an irreducible representation $\Gamma^{(l)}$ of $\boldsymbol{SO}(3)$ given by the matrices (3.2.34).

3.4 Spinor Representations

We are going to see that the irreducible representations, denoted $D^{(j)}$, of the group $\boldsymbol{SU}(2)$ also constitute irreducible representations for the rotation group $\boldsymbol{SO}(3)$. The j-spinors forming bases for the representations of $\boldsymbol{SU}(2)$, we shall say that we also have *spinor representations* of the group $\boldsymbol{SO}(3)$. We shall see that these representations include those given by spherical harmonics.

3.4.1 The Two-Dimensional Irreducible Representation

The Representation $\Gamma^{(1/2)}$ *in the Canonical Basis*

The irreducible representation $\Gamma^{(1/2)}$ of $\boldsymbol{SO}(3)$ of weight $j = \frac{1}{2}$ and dimension $n = 2j+1 = 2$ has as its infinitesimal matrices those given by (3.2.34), that is,

$$A_x = -\tfrac{1}{2}i \begin{bmatrix} 0 & 1 \\ 1 & 0 \end{bmatrix} = -\tfrac{1}{2}i\sigma_1, \tag{3.4.1a}$$

$$\begin{aligned} A_y &= -\tfrac{1}{2}i \begin{bmatrix} 0 & -i \\ i & 0 \end{bmatrix} = -\tfrac{1}{2}i\sigma_2, \\ A_z &= -\tfrac{1}{2}i \begin{bmatrix} 1 & 0 \\ 0 & -1 \end{bmatrix} = -\tfrac{1}{2}i\sigma_3. \end{aligned} \tag{3.4.1b}$$

The matrices $\sigma_1, \sigma_2, \sigma_3$ are the Pauli matrices which have already been introduced in the course of Chapter 1.

The Spinor Representation

The matrices (3.4.1) are exactly the rotation matrices (1.3.7), (1.3.9), (1.3.11) in the two-component spinor space which we obtained in Chapter 1. On the other hand, we have seen in Chapter 2 that these matrices form a two-dimensional irreducible representation $D^{(1/2)}$ of the group $\boldsymbol{SU}(2)$, the spinors (ψ, ϕ) of which form a basis of the representation.

The vector space of spinors thus also forms a representation space for the group $\boldsymbol{SO}(3)$. This representation is essentially spinorial, that is to say, there does not exist an equivalent representation of any other type, vector or tensor, which is not the case for three-dimensional representations and more generally for all those of odd dimension, as will be seen in the remainder.

The Binary Representation

Fundamentally, the coincidence of the irreducible representations of $\boldsymbol{SU}(2)$ and $\boldsymbol{SO}(3)$ come from a rotation of $\boldsymbol{SO}(3)$ corresponding to each transformation U of the group $\boldsymbol{SU}(2)$. Furthermore, the product R_2R_1 of two rotations corresponds to the product U_2U_1 of two rotations. However, the relation between $\boldsymbol{SU}(2)$ and $\boldsymbol{SO}(3)$ is not a rotation, since to a rotation R there correspond two transformations U and $-U$, as we have seen in Chapter 2. The groups $\boldsymbol{SU}(2)$ and $\boldsymbol{SO}(3)$ are called HOMOMORPHIC *groups*.

The matrices (3.4.1) or the representation $\Gamma^{(1/2)}$ thus have as their general expression the infinitesimal matrix (1.2.13). Let us write this latter matrix for a rotation through an angle θ about an axis along a unitary vector $\boldsymbol{n}$ with components n_x, n_y, n_z. We have

$$\begin{aligned} M(U) &= M[R(\boldsymbol{n}, \theta)] \\ &= \begin{bmatrix} \cos\frac{1}{2}\theta - in_z \sin\frac{1}{2}\theta & -(in_x + n_y)\sin\frac{1}{2}\theta \\ (-in_x + n_y)\sin\frac{1}{2}\theta & \cos\frac{1}{2}\theta + in_z\sin\frac{1}{2}\theta \end{bmatrix}. \end{aligned} \tag{3.4.2}$$

To the rotation $R(\boldsymbol{n}, \theta)$ there correspond two matrices $M(U)$ and $M(-U)$. The change of sign of the matrix elements (3.4.2) corresponds to the matrix

$$-M[R(\boldsymbol{n}, \theta)] = M[R(\boldsymbol{n}, \theta + 2\pi)].$$

It will be noticed that to a rotation through an angle $\theta = 2\pi$ there corresponds not the unit matrix I but $-I$.

As every rotation R of $\boldsymbol{SO}(3)$ can be represented by two matrices of the type of (3.4.2), we say that they form a BINARY REPRESENTATION.

3.4.2 The Three-Dimensional Irreducible Representation

If the representation $\Gamma^{(1/2)}$ is purely spinorial we are going to see that the representation $\Gamma^{(1)}$ of weight $j = 1$ is a spinor representation equivalent to other vector representations.

The Representation $\Gamma^{(1)}$ in the Canonical Basis

To the representation $\Gamma^{(1)}$ of weight $j = 1$ and dimension $n = 2j + 1 = 3$ there correspond the infinitesimal matrices (3.2.34) calculated from the vectors $\boldsymbol{x}_k$ of the canonical basis for $k = 1, 0, -1$, that is,

$$\begin{aligned} A_x &= -\frac{i}{\sqrt{2}} \begin{bmatrix} 0 & 1 & 0 \\ 1 & 0 & 1 \\ 0 & 1 & 0 \end{bmatrix}, \\ A_y &= -\frac{i}{\sqrt{2}} \begin{bmatrix} 0 & -i & 0 \\ i & 0 & -i \\ 0 & i & 0 \end{bmatrix}, \\ A_z &= -i \begin{bmatrix} 1 & 0 & 0 \\ 0 & 0 & 0 \\ 0 & 0 & -1 \end{bmatrix}, \end{aligned} \tag{3.4.3}$$

First of all let us show that this representation is equivalent to that obtained from the three-dimensional vector space E_3 the rotation matrices of which are given by (3.1.1) and (3.1.2).

The Equivalence Between $\Gamma^{(1)}$ and a Vector Representation

Let us write $\{\boldsymbol{e}_1, \boldsymbol{e}_2, \boldsymbol{e}_3\}$ for the orthonormal basis of the vector space E_3, and let $\{\boldsymbol{e}_1{}', \boldsymbol{e}_2{}', \boldsymbol{e}_3{}'\}$ be a new basis. Let us write P for the matrix taking the basis $\{\boldsymbol{e}_i\}$ to the basis $\{\boldsymbol{e}_j{}'\}$, that is to say in matrix form

$$[\boldsymbol{e}_j{}'] = P[\boldsymbol{e}_i]. \tag{3.4.4}$$

If (x, y, z) are the components of a vector $\boldsymbol{OM}$ of the space E_3 defined over the basis $\{\boldsymbol{e}_i\}$, then its components in the basis $\{\boldsymbol{e}_j{}'\}$ are given by

$$[\boldsymbol{x}'] = P^{-1}[\boldsymbol{x}], \tag{3.4.5}$$

where $[\boldsymbol{x}']$ and $[\boldsymbol{x}]$ are the column matrices formed by the components of the vector.

If $M(R)$ is a rotation matrix over the basis $\{\boldsymbol{e}_i\}$ then the expression $M'(R)$ of this matrix over the basis $\{\boldsymbol{e}_j{}'\}$ is

$$M'(R) = P^{-1}M(R)P. \tag{3.4.6}$$

The rotation matrices (3.1.1) and (3.1.2) have been calculated in an orthonormal basis $\{\boldsymbol{e}_1, \boldsymbol{e}_2, \boldsymbol{e}_3\}$. Let us consider the following change of basis defined by the alteration matrix

$$P = \frac{1}{\sqrt{2}} \begin{bmatrix} 1 & 0 & -1 \\ i & 0 & i \\ 0 & -\sqrt{2} & 0 \end{bmatrix}. \tag{3.4.7}$$

The inverse of this matrix allows us to obtain the new components x', y', z' of the vector $\boldsymbol{OM}$. We have

$$P^{-1} = \frac{1}{\sqrt{2}} \begin{bmatrix} 1 & -i & 0 \\ 0 & 0 & -\sqrt{2} \\ -1 & -i & 0 \end{bmatrix}, \tag{3.4.8}$$

whence

$$x' = \frac{1}{\sqrt{2}}(x - iy), \qquad y' = -z, \qquad z' = -\frac{1}{\sqrt{2}}(x + iy). \tag{3.4.9}$$

Let us consider the matrix $M[R_x(\alpha)]$ given by (3.1.1) and corresponding to a rotation through an angle α about the axis Ox. The expression for this matrix in the new basis is

$$\begin{aligned} M'[R_x(\alpha)] &= \frac{1}{\sqrt{2}} \begin{bmatrix} 1 & -i & 0 \\ 0 & 0 & -\sqrt{2} \\ -1 & -i & 0 \end{bmatrix} \begin{bmatrix} 1 & 0 & 0 \\ 0 & \cos\alpha & -\sin\alpha \\ 0 & \sin\alpha & \cos\alpha \end{bmatrix} \\ &\quad \times \frac{1}{\sqrt{2}} \begin{bmatrix} 1 & 0 & -1 \\ i & 0 & i \\ 0 & -\sqrt{2} & 0 \end{bmatrix} \\ &= \frac{1}{2} \begin{bmatrix} 1+\cos\alpha & -i\sqrt{2}\sin\alpha & -1+\cos\alpha \\ -i\sqrt{2}\sin\alpha & -2\cos\alpha & -i\sqrt{2}\sin\alpha \\ -1+\cos\alpha & -i\sqrt{2}\sin\alpha & 1+\cos\alpha \end{bmatrix}. \end{aligned} \tag{3.4.10}$$

Let us calculate the infinitesimal matrix deduced from $M'[R_x(\alpha)]$. We obtain

$$A_x{}' = -\frac{1}{\sqrt{2}} \begin{bmatrix} 0 & 1 & 0 \\ 1 & 0 & 1 \\ 0 & 1 & 0 \end{bmatrix}. \tag{3.4.11}$$

This is precisely the infinitesimal matrix A_x given by (3.4.3) for the irreducible representation $\Gamma^{(1)}$ of the groups $\boldsymbol{SO}(3)$. A straightforward calculation of the matrix $A_x{}'$ can evidently be realized by using the expression (3.1.4) for the infinitesimal matrix L_x, and we obtain

$$A_x{}' = P^{-1} L_x P = A_x. \tag{3.4.12}$$

It is easily verified that we obtain similarly the infinitesimal rotation matrices for the axes Oy and Oz

$$A_y = P^{-1} L_y P, \qquad A_z = P^{-1} L_z P. \tag{3.4.13}$$

The representation $\Gamma^{(1)}$ of $\boldsymbol{SO}(3)$ in the canonical basis is therefore equivalent to its vector representation which has the vector space E_3 as its representation space.

Spinor Equivalence

For the moment let us show that $\Gamma^{(i)}$ is a spinor representation equivalent to the three-dimensional irreducible representation $D^{(1)}$ of $\boldsymbol{SU}(2)$. According to the definition of the spinor (ψ, ϕ) given by relations (1.1.6) and (1.1.9), we have

$$x + iy = 2\psi^* \phi, \qquad x - iy = 2\psi\phi^*, \qquad z = \psi\psi^* - \phi\phi^*. \tag{3.4.14}$$

Now, the spinor $(\phi^*, -\psi^*)$ is transformed by the unitary transformation U given by (2.2.1) in the same way as the spinor (ψ, ϕ). In fact, we have

$$\begin{aligned} \begin{bmatrix} \psi' \\ \phi' \end{bmatrix} &= \begin{bmatrix} a & b \\ -b^* & a^* \end{bmatrix} \begin{bmatrix} \psi \\ \phi \end{bmatrix} \\ &= \begin{bmatrix} a\psi + b\phi \\ -b^*\psi + a^*\phi \end{bmatrix}, \end{aligned} \tag{3.4.15}$$

and on the other hand, taking account of relation (3.4.15),

$$\begin{aligned} \begin{bmatrix} a & b \\ -b^* & a^* \end{bmatrix} \begin{bmatrix} \phi^* \\ -\psi^* \end{bmatrix} &= \begin{bmatrix} a\phi^* - b\psi^* \\ -b^*\phi^* + a^*\psi^* \end{bmatrix} \\ &= \begin{bmatrix} \phi'^* \\ -\psi'^* \end{bmatrix}. \end{aligned} \tag{3.4.16}$$

This last relation shows that ϕ^* is transformed like ψ, and that $-\psi^*$ is transformed like ϕ. As a result the complex number given by (3.4.9) are transformed as follows

$$\begin{aligned} x' &= \frac{1}{\sqrt{2}}(x - iy) = \sqrt{2}\psi\phi^* \quad \text{transforms like } \sqrt{2}\psi^2, \\ y' &= -z = -(\psi\psi^* - \phi\phi^*) \quad \text{transforms like } 2\psi\phi, \\ z' &= -\frac{1}{\sqrt{2}}(x + iy) = -\sqrt{2}\psi^*\phi \quad \text{transforms like } \sqrt{2}\phi^2. \end{aligned}$$

Finally, the components x', y', z' are transformed like the components of the spinor of order $2j = 2$ given by relation (2.2.16). Consequently, the irreducible representation $D^{(1)}$ of the group $\boldsymbol{SU}(2)$ is equivalent to the representation $\Gamma^{(1)}$ of the group $\boldsymbol{SO}(3)$. Moreover, as $\Gamma^{(1)}$ is equivalent to a vector representation it does not form, oppositely to $\Gamma^{(1/2)}$, a purely spinor representation.

In order to prove that $\Gamma^{(1)}$ is a spinor representation we can also use the representation $D^{(1)}$ of $\boldsymbol{SU}(2)$, the matrix (2.2.19) of which allows us to obtain straightforwardly the infinitesimal matrices (3.4.3) of $\Gamma^{(1)}$. To do so it suffices to express the parameters a and b which appear in the matrix (2.2.18), by their form given by (3.4.2), that is,

$$a = \cos\tfrac{1}{2}\theta - in_z \sin\tfrac{1}{2}\theta, \qquad b = -(in_x + n_y)\sin\tfrac{1}{2}\theta. \tag{3.4.17}$$

We shall use this method of calculation for the $(2j+1)$-dimensional representations.

The Spherical Harmonics $Y_{lm}(\theta, \varphi)$

The spherical harmonics Y_{lm} for $m = -1, 0, 1$ also form a three-dimensional vector space giving a vector representation, denoted $F^{(1)}$, of the group $\boldsymbol{SO}(3)$. We are going to see that this latter representation is also equivalent to the spinor representation $D^{(1)}$ of $\boldsymbol{SU}(2)$. The functions $Y_{lm}(\theta, \varphi)$ are the following

$$Y_{10} = \sqrt{\frac{3}{4\pi}}\cos\theta, \qquad Y_{11} = -\sqrt{\frac{3}{8\pi}}\sin\theta e^{i\varphi}, \qquad Y_{1,-1} = \sqrt{\frac{3}{8\pi}}\sin\theta e^{-i\varphi}. \tag{3.4.18}$$

Let $\boldsymbol{OM}$ be a ray vector with components x, y, z. In spherical coordinates we have

$$x = r\sin\theta\cos\varphi, \qquad y = r\sin\theta\sin\varphi, \qquad z = r\cos\theta. \tag{3.4.19}$$

The spherical harmonics can thus be written with the help of Cartesian coordinates in the form

$$Y_{10} = \frac{1}{r}\sqrt{\frac{3}{4\pi}}z, \;\; Y_{11} = \frac{1}{r}\sqrt{\frac{3}{4\pi}}\frac{x+iy}{\sqrt{2}}, \;\; Y_{1,-1} = \frac{1}{r}\sqrt{\frac{3}{4\pi}}\frac{x-iy}{\sqrt{2}}. \tag{3.4.20}$$

We recover the form of expressions (3.4.9) for the change of basis which allows us to pass from the vector representation to the spinor representation $D^{(1)}$. As the norm r of the vector $\boldsymbol{OM}$ is invariant under rotation, the spherical harmonics Y_{1m} thus form, to within numerical factors, the components of vectors of the representation space of $D^{(1)}$. The space of spherical harmonics Y_{1m} and that of the spinors $(\psi^2, \psi\phi, \phi^2)$ are thus the spaces of two equivalent representations of the group $\boldsymbol{SO}(3)$ of rotations.

3.4.3 $(2j+1)$-Dimensional Irreducible Representations

Quite generally, all the irreducible representations of $\boldsymbol{SU}(2)$ are equivalent to irreducible representations of $\boldsymbol{SO}(3)$. This results, as we have seen beforehand, from the homomorphism which exists between these two groups. We have established these equivalences straightforwardly in the case of two- and three-dimensional representations. Let us return first to a direct proof of the equivalence between the representation $D^{(1)}$ of $\boldsymbol{SU}(2)$ and $\Gamma^{(1)}$ of $\boldsymbol{SO}(3)$.

Three-Dimensional Equivalent Representations

Let us consider the matrix (2.2.19) in which we are going to express the parameters a and b by their expression (3.4.17) already given. In order to simplify the general expression of this matrix let us simply study a rotation through an angle θ about the axis Ox, for which we have $n_x = 1$, $n_y = n_z = 0$, whence the expression for the parameters

$$a = \cos\tfrac{1}{2}\theta, \qquad b = -i\sin\tfrac{1}{2}\theta. \tag{3.4.21}$$

We then obtain the following rotation matrix in spinor space

$$M^{(1)}[R_x(\theta,0,0)] = \begin{bmatrix} \cos^2\frac{1}{2}\theta & -i\sqrt{2}\cos\frac{1}{2}\theta\sin\frac{1}{2}\theta & \sin^2\frac{1}{2}\theta \\ -i\sqrt{2}\cos\frac{1}{2}\theta\sin\frac{1}{2}\theta & \cos^2\frac{1}{2}\theta - \sin^2\frac{1}{2}\theta & -i\sqrt{2}\cos\frac{1}{2}\theta\sin\frac{1}{2}\theta \\ -\sin^2\frac{1}{2}\theta & -i\sqrt{2}\cos\frac{1}{2}\theta\sin\frac{1}{2}\theta & \cos^2\frac{1}{2}\theta \end{bmatrix}. \tag{3.4.22}$$

Let us calculate the derivative of each matrix element with respect to θ for $\theta = 0$. We easily obtain the matrix A_x given by (3.4.3). Calculating similarly the matrices $M^{(1)}[R_y(0,\theta,0)]$ and $M^{(1)}[R_z(0,0,\theta)]$ respectively corresponding to the rotations about the axes Oy and Oz, we obtain infinitesimal matrices identical to the matrices A_y and A_z given by (3.4.3). This directly shows the equivalence between the spinor representation $D^{(1)}$ of $\boldsymbol{SU}(2)$ and $\Gamma^{(1)}$ of $\boldsymbol{SO}(3)$.

$(2j+1)$-Dimensional Representations

We can generalize the previous calculation to a representation $D^{(j)}$ of $\boldsymbol{SU}(2)$, the matrix elements of which are given by expression (2.2.27), that is to say,

$$U^{(j)}{}_{m'm}(a,b) = \sum_{\lambda} \frac{[(j+m)!(j-m)!/,(j+m')!(j-m')!/,]^{1/2}}{(j+m-\lambda)!\lambda!(j-m'-\lambda)!(m'-m+\lambda)!}$$

$$\times a^{j+m-\lambda}(a^*)^{j-m'-\lambda}b^{\lambda}(-b^*)^{m'-m+\lambda}. \tag{3.4.23}$$

Changing a and b according to their expression by the functions of (3.4.17) for each of the particular rotations about the axes Ox, Oy, Oz allows us to calculate the matrices $M^{(j)}(R)$ and then the corresponding infinitesimal matrices. Let us make this calculation for the rotation about the Oz axis for which $n_z = 1$, $n_x = n_y = 0$, whence

$$a = \cos\tfrac{1}{2}\theta - i\sin\tfrac{1}{2}\theta = e^{-i\theta/2}, \qquad b = 0. \tag{3.4.24}$$

The nonzero matrix elements of (3.4.23) are those in which b does not appear, and they correspond to $m = m'$ and $\lambda = 0$. Thus we obtain a diagonal matrix which has as its general term

$$U^{(j)}{}_{m'm} = e^{i\theta(m+m')/2}\delta_{mm'}. \tag{3.4.25}$$

The corresponding infinitesimal matrix thus has as its elements

$$(A_z)_{mm'} = -\tfrac{1}{2}i(m+m')\delta_{mm'}. \tag{3.4.26}$$

Since m and m' take values from j to $-j$ we recover the expression for the diagonal matrix A_z of the representation $\Gamma^{(1)}$ given by (3.2.34).

Finally, all the irreducible representations $D^{(j)}$ of $\boldsymbol{SU}(2)$ are thus spinor representations of $\boldsymbol{SO}(3)$ equivalent to the irreducible representations $\Gamma^{(j)}$ of this group. The representations given by the spherical harmonics correspond to the integer values j; these are the odd-dimensional representations. Those corresponding to even-dimensional half-integer values of j are pure spinor representations having no equivalent in the vector space of differential functions $f(\theta,\varphi)$.

Representations in Functions of the Euler Angles

We often use the Euler angles α, β, γ defined in Chapter 1 as parameters of $\boldsymbol{SO}(3)$. The matrix (1.2.29) then gives us as an expression for a and b

$$a = e^{-i(\alpha+\gamma)/2}\cos\tfrac{1}{2}\beta, \qquad b = -ie^{-i(\alpha-\gamma)/2}\sin\tfrac{1}{2}\beta. \tag{3.4.27}$$

Substituting expressions (3.4.22) into relation (2.2.27) we obtain the matrix elements of the irreducible representations $\Gamma^{(j)}$ of $\boldsymbol{SO}(3)$ in the form

$$U^{(j)}{}_{qm} = \sum_{k=0}^{\infty} \frac{\sqrt{(j+m)!(j-m)!(j+q)!(j-q)!}}{k!(j+m-k)!(j-q-k)!(q-m+k)!} \times (-i)^{2k+q-m} e^{-i(m\alpha+q\gamma)} \left(\cos \tfrac{1}{2}\beta\right)^{2j+m-q-2k} \left(\sin \tfrac{1}{2}\beta\right)^{2k+q-m}, \tag{3.4.28}$$

with $q, m = -j, -j+1, \ldots, j-1, j$, $j = \frac{1}{2}, 1, \frac{3}{2}, 2, \ldots$. Each sum consists of only a finite number of terms since $1/n! = 0$ if $n < 0$. The representation $\Gamma^{(j)}$ is called a *binary representation* for the half-integer values of j and a *unary representation* for the integer values.

3.5 Solved Problems

EXERCISE **3.1**

1. Consider a rotation about the axis Ox through an angle α. Calculate the matrix $M^{(1/2)}[R(\alpha, 0, 0)]$ of the two-dimensional spinor representation of the group $\boldsymbol{SO}(3)$ starting from the infinitesimal matrix $A_x^{(1/2)}$ given by (3.4.1).

2. The same question for rotations about the axes Oy and Oz.

3. Show that the three matrices are particular cases of the matrix $M[R(\boldsymbol{n}, \theta)]$ given by (3.4.2).

SOLUTION **3.1**

1. The matrix $M^{(1/2)}[R(\alpha, 0, 0)]$ with respect to a rotation through an angle α about the axis Ox is obtained by solving a differential system analogous to that given by (2.1.47). Thus in the present case,

$$\frac{\mathrm{d}R}{\mathrm{d}\beta} = A_x{}^{(1/2)} R,$$

with $R = M^{(1/2)}[R(\alpha, 0, 0)]$ and the initial condition $M^{(1/2)}[R(0, 0, 0)] = \mathbb{1}$. The infinitesimal matrix $A_x{}^{(1/2)}$ with respect to the representation $\Gamma^{(1/2)}$ is given by formula (3.4.1), that is,

$$A_x{}^{(1/2)} = -\frac{i}{2} \begin{bmatrix} 0 & 1 \\ 1 & 0 \end{bmatrix}.$$

Let us write R_{ij} for the matrix elements of R. The differential system is written

$$\frac{\mathrm{d}R_{11}}{\mathrm{d}\beta} = -\frac{i}{2}R_{21}, \qquad \frac{\mathrm{d}R_{12}}{\mathrm{d}\beta} = -\frac{i}{2}R_{22},$$
$$\frac{\mathrm{d}R_{21}}{\mathrm{d}\beta} = -\frac{i}{2}R_{11}, \qquad \frac{\mathrm{d}R_{22}}{\mathrm{d}\beta} = -\frac{i}{2}R_{12},$$

with the initial conditions resulting from $M^{(1/2)}[R(0,0,0)] = \mathbb{1}$

$$R_{11}(0) = R_{22}(0) = 1, \qquad R_{12}(0) = R_{21}(0) = 0.$$

The solution of the differential system satisfying the initial conditions is

$$M^{(1/2)}[R(\alpha,0,0)] = \begin{bmatrix} \cos\frac{1}{2}\alpha & -i\sin\frac{1}{2}\alpha \\ -i\sin\frac{1}{2}\alpha & \cos\frac{1}{2}\alpha \end{bmatrix}.$$

2. The infinitesimal matrices ${A_y}^{(1/2)}$ and ${A_z}^{(1/2)}$ given by (3.4.1) allow us to obtain by an analogous calculation the matrices corresponding to the rotations about the axes Oy and Oz through respective angles β and γ, that is,

$$M^{(1/2)}[R(0,\beta,0)] = \begin{bmatrix} \cos\frac{1}{2}\beta & -\sin\frac{1}{2}\beta \\ \sin\frac{1}{2}\beta & \cos\frac{1}{2}\beta \end{bmatrix},$$

$$M^{(1/2)}[R(0,0,\gamma)] = \begin{bmatrix} e^{-i\gamma/2} & 0 \\ 0 & e^{i\gamma/2} \end{bmatrix}.$$

3. Formula (3.4.2) has as its expression

$$\begin{aligned} M(U) &= M[R(\boldsymbol{n},\theta)] \\ &= \begin{bmatrix} \cos\frac{1}{2}\theta - in_z\sin\frac{1}{2}\theta & -(in_x+n_y)\sin\frac{1}{2}\theta \\ (-in_x+n_y)\sin\frac{1}{2}\theta & \cos\frac{1}{2}\theta + in_z\sin\frac{1}{2}\theta \end{bmatrix}. \end{aligned}$$

The components n_x, n_y, n_z of the unitary vector along the axis of rotation is reduced to $n_x = 1$ and $n_y = n_z = 0$ for a rotation about the axis Ox. The matrix $M[R(\boldsymbol{n},\theta)]$ gives the matrix $M^{(1/2)}[R(\alpha,0,0)]$ for these values of the components. Analogous remarks allow us to deduce the matrices $M^{(1/2)}[R(0,\beta,0)]$ and $M^{(1/2)}[R(0,0,\gamma)]$ from $M[R(\boldsymbol{n},\theta)]$.

Exercise **3.2**

The three-dimensional irreducible representation of the group $\boldsymbol{SU}(2)$ is given by the matrix (2.2.19). Show that the three-dimensional infinitesimal matrix of the irreducible representation is identical to the three-dimensional infinitesimal matrix in the canonical basis of the group $\boldsymbol{SO}(3)$ given by (3.2.34).

Solution **3.2**

The expressions for the parameters a and b which appear in the matrix (2.2.19) are given by the formulas (1.2.13), that is,

$$a = \cos\tfrac{1}{2}\theta - iL_3 \sin\tfrac{1}{2}\theta, \qquad b = -(iL_1 + L_2)\sin\tfrac{1}{2}\theta,$$

where θ is the angle of rotation about an axis along the unitary vector with components L_1, L_2, L_3. For a rotation about the axis Oz we have $L_3 = 1$, $L_1 = L_2 = 0$, whence

$$a = \cos\tfrac{1}{2}\theta - i\sin\tfrac{1}{2}\theta, \qquad b = 0.$$

With these values the matrix (2.2.19) becomes

$$M_z^{(1)}(U) = \begin{bmatrix} a^2 & 0 & 0 \\ 0 & aa^* & 0 \\ 0 & 0 & a^{*2} \end{bmatrix}.$$

Using the value of a and then differentiating the matrix elements with respect to θ and taking the value of these derivatives at $\theta = 0$, we obtain

$$A^{(1)}(U) = -i\begin{bmatrix} 1 & 0 & 0 \\ 0 & 0 & 0 \\ 0 & 0 & -1 \end{bmatrix}.$$

This is the three-dimensional matrix of the group $\boldsymbol{SU}(2)$ which is identical to that of the group $\boldsymbol{SO}(3)$ given by (3.2.34).

Exercise **3.3**

Let us study the rotation $R(\varphi)$ through an angle φ of the vector $\boldsymbol{u}$ about the axis Oz of a Cartesian frame of reference $Oxyz$. The origin of the vector $\boldsymbol{u}$ is located on the axis Oz. We write the components of $\boldsymbol{u}$ as u_x, u_y, u_z.

1. Calculate the components $u_x{}', u_y{}', u_z{}'$ of the vector $\boldsymbol{u}$ when rotated.

2. Write the rotation matrix $M[R(\varphi)]$.

SOLUTION **3.3**

1. Let us consider a rotation in the direction which goes from the axis Ox to Oy. Let us write $\boldsymbol{e}_1, \boldsymbol{e}_2, \boldsymbol{e}_3$ for the orthonormal vectors of the Cartesian frame of reference. We have

$$\boldsymbol{u} = u_x \boldsymbol{e}_1 + u_y \boldsymbol{e}_2 + u_z \boldsymbol{e}_3.$$

Let Γ be the operator associated with the rotation $R(\varphi)$. This yields

$$\begin{aligned} \boldsymbol{u}' = \Gamma(\boldsymbol{u}) &= u_x \Gamma(\boldsymbol{e}_1) + u_y \Gamma(\boldsymbol{e}_2) + u_z \Gamma(\boldsymbol{e}_3) \\ &= u_x{}' \boldsymbol{e}_1 + u_y{}' \boldsymbol{e}_2 + u_z{}' \boldsymbol{e}_3. \end{aligned}$$

A rotation through an angle φ about the Oz axis leaves $\boldsymbol{e}_3$ unchanged, whence $\Gamma(\boldsymbol{e}_3) = \boldsymbol{e}_3$. The rotation of the vectors $\boldsymbol{e}_2$ and $\boldsymbol{e}_3$ takes place in the plane xOy. The components of the transformed vectors are given by their projections on the axes Ox and Oy, whence

$$\begin{aligned} \Gamma(\boldsymbol{e}_1) &= \cos\varphi \boldsymbol{e}_1 + \sin\varphi \boldsymbol{e}_2, \\ \Gamma(\boldsymbol{e}_2) &= -\sin\varphi \boldsymbol{e}_1 + \cos \boldsymbol{e}_2. \end{aligned}$$

By identification with the expanded expression for $\boldsymbol{u}'$ this yields

$$\begin{aligned} u_x{}' &= u_x \cos\theta - u_y \sin\varphi, \\ u_y{}' &= u_x \sin\varphi + u_y \cos\varphi, \\ u_z{}' &= u_z. \end{aligned}$$

2. The change of components may be written in the matrix form

$$\begin{aligned} \begin{bmatrix} u_x{}' \\ u_y{}' \\ u_z{}' \end{bmatrix} &= M[C(\varphi)] \begin{bmatrix} u_x \\ u_y \\ u_z \end{bmatrix} \\ &= \begin{bmatrix} \cos\varphi & -\sin\varphi & 0 \\ \sin\varphi & \cos\varphi & 0 \\ 0 & 0 & 1 \end{bmatrix} \begin{bmatrix} u_x \\ u_y \\ u_z \end{bmatrix}. \end{aligned}$$

EXERCISE **3.4**

1. Recall the meaning of the antisymmetric Kronecker symbol ε_{ijk}.
2. Show the validity of formuals (3.1.6)

$$(L_k)_{ij} = -\varepsilon_{ijk}, \qquad i, j, k = 1, 2, 3.$$

3. Use the Kronecker symbol to write a formula for the commutation relations of the infinitesimal matrices given by (3.1.7).

SOLUTION **3.4**

1. The Kronecker symbol ε_{ijk} for $i, j, k = 1, 2, 3$ is a tensor [10] which has the following values:

   — $\varepsilon_{ijk} = 0$ if any two indices have the same value. For example, $\varepsilon_{112} = \varepsilon_{313} = 0$;

   — $\varepsilon_{ijk} = 1$ if the indices are in the initial order 1,2,3 or result from an even number of permutations of the intial order. For example, $\varepsilon_{123} = \varepsilon_{231} = 1$;

   — $\varepsilon_{ijk} = -1$ if the indices are in an order resulting from an odd number of permutations of the initial order. For example, $\varepsilon_{132} = \varepsilon_{231} = -1$.

2. Let us consider the matrix $L_1 = L_x$ given by (3.1.4). We easily verify that

$$(L_1)_{ij} = -\varepsilon_{ij1}.$$

   We verify similarly

$$\begin{aligned} (L_2)_{ij} &= -\varepsilon_{ij2} \qquad \text{for} \quad L_2 = L_y, \\ (L_3)_{ij} &= -\varepsilon_{ij3} \qquad \text{for} \quad L_3 = L_z. \end{aligned}$$

   These three formulas lead to the required relation

$$(L_k)_{ij} = -\varepsilon_{ijk}, \qquad i, j, k = 1, 2, 3.$$

3. Using the preceding notations relations (3.1.7) can be written in the form

$$[L_i, L_j] = \sum_{k=1}^{3} \varepsilon_{ijk} L_k.$$

   We verify, for example, that

$$\begin{aligned} [L_1, L_2] &= \varepsilon_{121} L_1 + \varepsilon_{122} L_2 + \varepsilon_{123} L_3 \\ &= L_3. \end{aligned}$$

4

Pauli Spinors

4.1 Spin and Spinors

The first notion of spin arose from the experimental data on the anomalous Zeeman effect. To explain the results obtained, Uhlenbeck and Goudsmit had the idea in 1925 of attributing to the electron its own magnetic moment, or *spin*. Thus the anomalous Zeeman effect began to receive an explanation, thanks to this hypothesis which, however, was still in the framework of classical electromagnetism.

Next, in 1927, Pauli's theory made the first attempt at accurately setting the problem of the magnetic electron in the framework of the general ideas of quantum mechanics. According to the principles of the latter we cannot attribute a definite spin to an individual electron but can only speak of the probability of an experiment which allows the direction of the spin of an electron to be determined giving such and such a result. Since the component of the spin has two opposite values along an axis, it is natural to assume that we must associate not one function ψ but two functions ψ_1 and ψ_2 with an electron, these functions being such that they must allow us to define the probability of the possible orientations of the spin in a certain direction.

In addition to these fundamental ideas Pauli established two equations satisfied by the functions ψ_1 and ψ_2, equations analogous to the Schrödinger equation but in which the magnetic moment of the electron appears. Under a rotation of the coordinate axes we show that the Pauli equations keep the same form, but that the functions ψ_1 and ψ_2 are transformed into each other in a way which corresponds exactly with the *spinor transformation formula*.

It was Pauli's theory, and then Darwin's—which sought to introduce the electron's magnetism in a way conforming to the theory of relativity by defining four functions representing the components of a space–time vector—which inspired Dirac to invent his theory of the relativistic electron.

Dirac, examining the previous relativistic theories, was led to a new hypothesis, that the equations controlling the evolution of the components ψ_i of the wave function must be first order with respect to the four variables x, y, z, t, although the relativistic equations generalizing Schrödinger's equation were second order in these variables.

The idea of linearizing Schrödinger's equation itself was developed next, inspiring Dirac's works, and allowed the Pauli equations to be recovered in which the spin was then introduced automatically. The existence of spin is consequently not a purely relativistic effect, but becomes a consequence of the linearization of the wave equations. We are going to establish these linearized equations from the paper by Lévy-Leblond [15].

4.2 The Linearized Schrödinger Equations

4.2.1 The Free Particle

The classical Schrödinger equation is a partial differential equation of the first order with respect to time and second order with respect to the spacial variables. Let us seek a linear equation equivalent to the Schrödinger equation having derivatives entirely of the first order with respect to all the variables.

Let us write $\boldsymbol{P}$ for the momentum vector's operator with the operators P_i, $i = 1, 2, 3$, as its Cartesian components. E is the operator $i\hbar\partial/\partial t$. Let us look for a wave equation for a free quantum particle in the form

$$H_1\Psi = (AE + \boldsymbol{B}\cdot\boldsymbol{P} + C)\Psi = 0, \tag{4.2.1}$$

where $A, \boldsymbol{B}, C$ are linear operators to be determined. $\boldsymbol{B}$ is a vector operator with components B_1, B_2, B_3.

Let us seek the solutions Ψ of (4.2.1) which simultaneously satisfy the Schrödinger equation

$$(H - E)\Psi = \left(\frac{\boldsymbol{P}^2}{2m} - E\right)\Psi = 0, \tag{4.2.2}$$

where H is the Hamiltonian operator of the system. Since (4.2.4) consists of differential operators of the second order it will be possible to obtain them by multiplying the operators of (4.2.1) by another operator of the form

$$H_2 = A'E + \boldsymbol{B}'\cdot\boldsymbol{P} + C', \tag{4.2.3}$$

where A', $\boldsymbol{B}'$, and C are operators to be determined, and such that

$$H_2 H_1 = k(H - E), \tag{4.2.4}$$

where k is an arbitrary constant which we shall set equal to $-2m$.

The Determination of the Linear Operators

Let us expand (4.2.4). We obtain

$$(A'E + B_i{}'P_i + C')(AE + B_jP_j + C) = 2mE - P_kP_k,$$

where the summation is carried out according to the Einstein convention. Identifying sides we obtain the following relations

$$A'A = 0, \qquad A'C + C'A = 2m, \qquad C'C = 0, \qquad A'B_j + B_j{}'A = 0,$$
$$B_i{}'B_j + B_j{}'B_i = -2\delta_{ij}, \qquad C'B_i + B_i{}'C = 0. \tag{4.2.5}$$

In order to write these conditions in condensed form, let us introduce the following notations

$$B_4 = i\left(A + \frac{1}{2m}C\right), \quad B_4{}' = i\left(A' + \frac{1}{2m}C'\right),$$
$$B_5 = A - \frac{1}{2m}C, \qquad B_5{}' = A' - \frac{1}{2m}C'. \tag{4.2.6}$$

Conditions (4.2.5) may then be written in the condensed form

$$B_i{}'B_j + B_j{}'B_i = -2\delta_{ij} \qquad i,j = 1,\ldots,5. \tag{4.2.7}$$

In order to recover the classical matrices of relativistic quantum mechanics let us make a new change of operators. To this end let us consider an arbitrary new operator M such that M^{-1} exists. Let us choose M such that

$$B_5 = -iM, \quad B_5{}' = -iM^{-1}, \quad B_k = M\gamma_k, \quad B_k{}' = -\gamma_k M^{-1}, \tag{4.2.8}$$

with $k = 1, 2, 3, 4$. Let us introduce expressions (4.2.8) into relations (4.2.7). We obtain the following *anticommutation relations*

$$\gamma_j\gamma_k + \gamma_k\gamma_j = 2\delta_{jk}, \qquad j,k = 1,2,3,4. \tag{4.2.9}$$

Although there are no more than four indices for the operators γ_k it can be remarked that relations (4.2.7) are automatically satisfied when one of their indices i, j is equal to 5 or if $i = j = 5$. In fact

$$B_5{}'B_j + B_j{}'B_5 = -iM^{-1}M\gamma_j + \gamma_j M^{-1}iM = 0, \qquad j = 1,\ldots,4,$$
$$B_5{}'B_5 + B_5{}'B_5 = 2iM^{-1}iM = -2. \tag{4.2.10}$$

Matrices of Operators

The operators B_i are finally replaced by the four operators γ_j and the operator M. Relations (4.2.9) show that we have

$$(\gamma_1)^2 = (\gamma_2)^2 = (\gamma_3)^2 = (\gamma_4)^2 = 1. \tag{4.2.11}$$

The operators γ_j thus have eigenvalues ± 1, and the matrices which represent them have a square equal to the unit matrix. According to (4.2.9) the operators γ_j anticommute

$$\gamma_i \gamma_j = -\gamma_j \gamma_i \qquad \text{for} \quad i \neq j. \tag{4.2.12}$$

From this latter relation we deduce that the trace of the matrices of these operators is zero. In fact, relations (4.2.11) and (4.2.12) give us

$$\gamma_j = -\gamma_i \gamma_j \gamma_i, \tag{4.2.13}$$

whence, by also writing γ_i for the matrix of the operator γ_i and taking account of the trace of a product of matrices being independent of the order of the product

$$\operatorname{Tr} \gamma_j = -\operatorname{Tr} \gamma_i \gamma_j \gamma_i = -\operatorname{Tr} \gamma_j \gamma_i^2 = -\operatorname{Tr} \gamma_j. \tag{4.2.14}$$

Thus we obtain a trace of the matrices γ_j which is zero. Since the trace of a matrix is equal to the sum of its eigenvalues, and since the latter are equal to ± 1, the number of positive and negative values of the eigenvalues must be equal. Consequently the order of the matrices must be even. Let us seek matrices of the smallest possible order.

The matrices of order 2 cannot be admitted. In fact, they form a four-dimensional vector space, and we have seen in Chapter 1 that the unit matrix and the three Pauli matrices are able to form a basis of this vector space. If the Pauli matrices satisfy the anticommutation relations (4.2.12) they are only three in number, whilst four matrices γ_j are needed.

The matrices of order 2 are going to allow us to satisfy the conditions required. In order to obtain the anticommutative matrices let us use the Pauli matrices σ_i to obtain the matrices γ_j, that is,

$$\gamma_j = \begin{bmatrix} 0 & \sigma_j \\ \sigma_j & 0 \end{bmatrix} \quad \text{with } j = 1, 2, 3, \qquad \gamma_4 = \begin{bmatrix} \mathbb{1} & 0 \\ 0 & -\mathbb{1} \end{bmatrix}, \tag{4.2.15}$$

where $\mathbb{1}$ is the unit matrix of order 2. These 4×4 matrices satisfy relations (4.2.11) and (4.2.12), thus we can do calculations easily by using the properties of the Pauli matrices. We obtain

$$\gamma_i \gamma_j + \gamma_j \gamma_i = 2\delta_{ij} \begin{bmatrix} \mathbb{1} & 0 \\ 0 & \mathbb{1} \end{bmatrix}, \quad i, j = 1, 2, 3, 4. \tag{4.2.16}$$

Calculation of the matrices of the operators B_k requires the choice of the matrix of the operator M to allow relations (4.2.8) to be satisfied. In particular, it is necessary that M be invertible. Let us choose the matrix

$$M = \begin{bmatrix} 0 & 1\!\!1 \\ 1\!\!1 & 0 \end{bmatrix}, \tag{4.2.17}$$

which gives us for the matrices of the operators B_j,

$$B_j = M\gamma_j = \begin{bmatrix} \sigma_j & 0 \\ 0 & \sigma_j \end{bmatrix}, \quad j = 1, 2, 3, \tag{4.2.18}$$

$$B_4 = M\gamma_4 = \begin{bmatrix} 0 & -1\!\!1 \\ 1\!\!1 & 0 \end{bmatrix}, \quad B_5 = -iM = -i\begin{bmatrix} 0 & 1\!\!1 \\ 1\!\!1 & 0 \end{bmatrix}. \tag{4.2.19}$$

Equation (4.2.1) sought also introduces the operators A and C which are related to B_4 and B_5 by the following relations deduced from (4.2.6)

$$A = \tfrac{1}{2}(B_5 - iB_4), \qquad C = -m(B_5 + iB_4). \tag{4.2.20}$$

The matrices (4.2.19) then give us for the matrices of the operators A and C

$$A = -i\begin{bmatrix} 0 & 0 \\ 1\!\!1 & 0 \end{bmatrix}, \qquad C = 2mi\begin{bmatrix} 0 & 1\!\!1 \\ 0 & 0 \end{bmatrix}. \tag{4.2.21}$$

As the wave equation operator (4.2.1) is described by fourth-order matrices the wave function Ψ of this equation must have four components, which we write

$$\Psi = \begin{bmatrix} \psi_1 \\ \psi_2 \\ \eta_1 \\ \eta_2 \end{bmatrix} = \begin{bmatrix} \psi \\ \eta \end{bmatrix} \quad \text{with} \quad \psi = \begin{bmatrix} \psi_1 \\ \psi_2 \end{bmatrix}, \quad \eta = \begin{bmatrix} \eta_1 \\ \eta_2 \end{bmatrix}. \tag{4.2.22}$$

Let us write $\boldsymbol{\sigma}$ for the vector which has as its components the Pauli matrices $\sigma_1, \sigma_2, \sigma_3$. We can then write the matrix of the vector operator $\boldsymbol{B}$, taking into account expressions (4.2.18) for the operators B_i forming the components of $\boldsymbol{B}$, in the form

$$\boldsymbol{B} = \begin{bmatrix} \boldsymbol{\sigma} & 0 \\ 0 & \boldsymbol{\sigma} \end{bmatrix}. \tag{4.2.23}$$

Substituting the expressions for the matrices $A, \boldsymbol{B}, C$ into (4.2.1) we obtain the linearized wave equation in the form

$$\left\{-i\begin{bmatrix} 0 & 0 \\ 1\!\!1 & 0 \end{bmatrix} E + \begin{bmatrix} \boldsymbol{\sigma} & 0 \\ 0 & \boldsymbol{\sigma} \end{bmatrix} \cdot \boldsymbol{P} + 2im\begin{bmatrix} 0 & 1\!\!1 \\ 0 & 0 \end{bmatrix}\right\}\begin{bmatrix} \psi \\ \eta \end{bmatrix} = 0. \tag{4.2.24}$$

The latter equation can be decomposed into two equations making use of the second-order matrices, that is,

$$\boldsymbol{\sigma}\cdot\boldsymbol{P}\psi + 2im\eta = 0, \qquad \boldsymbol{\sigma}\cdot\boldsymbol{P}\eta - iE\psi = 0. \tag{4.2.25}$$

4.2.2 Particle in an Electromagnetic Field

Equations (4.2.25) have been established for a free particle. If, for the moment, we consider a charged particle placed in an electromagnetic field it is necessary to make the following substitutions for the operators:

— for the operator $i\hbar\,\partial/\partial t$ we substitute the operator $i\hbar\,\partial/\partial t - eV$, where e is the electric charge of the particle and V is the electric potential;

— for the operator $-i\hbar\,\boldsymbol{\nabla}$ we substitute the operator $-i\hbar\,\boldsymbol{\nabla} - (e/c)\boldsymbol{A}$, where c is the speed of light and $\boldsymbol{A}$ is the vector potential of the electromagnetic field.

With these substitutions (4.2.25) becomes

$$\boldsymbol{\sigma}\cdot\left(\boldsymbol{P} - \frac{e}{c}\boldsymbol{A}\right)\psi + 2im\eta = 0, \tag{4.2.26}$$

$$\boldsymbol{\sigma}\cdot\left(\boldsymbol{P} - \frac{e}{c}\boldsymbol{A}\right)\eta - i(E - eV)\psi = 0. \tag{4.2.27}$$

Pauli's Equation

Eliminating the function η will allow us to obtain Pauli's equation for a particle in an electromagnetic field. Equation (4.2.26) gives us for the expression for η

$$\eta = \frac{i}{2m}\boldsymbol{\sigma}\cdot\left(\boldsymbol{P} - \frac{e}{c}\boldsymbol{A}\right)\psi. \tag{4.2.28}$$

Substituting this expression for η in (4.2.27) we obtain

$$\left\{E - eV - \frac{1}{2m}\left[\boldsymbol{\sigma}\cdot\left(\boldsymbol{P} - \frac{e}{c}\boldsymbol{A}\right)\right]\left[\boldsymbol{\sigma}\cdot\left(\boldsymbol{P} - \frac{e}{c}\boldsymbol{A}\right)\right]\right\}\psi = 0. \tag{4.2.29}$$

On the other hand, we have the following classical identity, where $\boldsymbol{B}$ and $\boldsymbol{C}$ are arbitrary vectors,

$$(\boldsymbol{\sigma}\cdot\boldsymbol{B})(\boldsymbol{\sigma}\cdot\boldsymbol{C}) = \boldsymbol{B}\cdot\boldsymbol{C} + i\boldsymbol{\sigma}\cdot(\boldsymbol{B}\times\boldsymbol{C}). \tag{4.2.30}$$

In the case where $\boldsymbol{B} = \boldsymbol{C} = \boldsymbol{P} - (e/c)\boldsymbol{A}$ relation (4.2.30) may be written

$$\left[\boldsymbol{\sigma}\cdot\left(\boldsymbol{P} - \frac{e}{c}\boldsymbol{A}\right)\right]\left[\boldsymbol{\sigma}\cdot\left(\boldsymbol{P} - \frac{e}{c}\boldsymbol{A}\right)\right]$$

$$= \left(\boldsymbol{P} - \frac{e}{c}\boldsymbol{A}\right)^2 + i\boldsymbol{\sigma}\cdot\left[\left(\boldsymbol{P} - \frac{e}{c}\boldsymbol{A}\right)\times\left(\boldsymbol{P} - \frac{e}{c}\boldsymbol{A}\right)\right]. \tag{4.2.31}$$

The vector product which appears in the right-hand side of (4.2.31) gives us, after expansion,

$$\left(\boldsymbol{P} - \frac{e}{c}\boldsymbol{A}\right) \times \left(\boldsymbol{P} - \frac{e}{c}\boldsymbol{A}\right) = -\frac{e}{c}(\boldsymbol{P} \times \boldsymbol{A}). \tag{4.2.32}$$

Equation (4.2.29) then becomes, on taking account of (4.2.31) and (4.2.32),

$$\left\{E - eV - \frac{1}{2m}\left(\boldsymbol{P} - \frac{e}{c}\boldsymbol{A}\right)^2 + \frac{ie}{2mc}\boldsymbol{\sigma}\cdot(\boldsymbol{P} \times \boldsymbol{A})\right\}\psi = 0. \tag{4.2.33}$$

Let us write $\boldsymbol{B}$ for the magnetic induction vector, then we have

$$\boldsymbol{P} \times \boldsymbol{A} = -i\hbar \boldsymbol{\nabla} \times \boldsymbol{A} = -i\hbar \boldsymbol{B}. \tag{4.2.34}$$

Equation (4.2.33) may then be written

$$\left\{E - eV - \frac{1}{2m}\left(\boldsymbol{P} - \frac{e}{c}\boldsymbol{A}\right)^2 + \frac{e\hbar}{2mc}\boldsymbol{\sigma}\cdot\boldsymbol{B}\right\}\psi = 0. \tag{4.2.35}$$

Thus we obtain the Pauli equation in which the term $(e\hbar/2mc)\boldsymbol{\sigma}\cdot\boldsymbol{B}$ appears, and which represents the interaction energy of the magnetic field with the intrinsic magnetic moment of the electron. Although Pauli had added this term in Schrödinger's equation in such a way as to make the theoretical and experimental results agree, we see that the spin is here introduced automatically as a consequence of the postulate of linearization of the wave equation. Moreover, this latter theory gives the correct value of the intrinsic magnetic moment of the electron.

4.2.3 The Spinors in Pauli's Equation

Under a rotation of the coordinate system the wave function ψ which appears in Pauli's equation (4.2.35) transforms like a spinor. Let us study a particular example which allows us to calculate the transformation matrix of the spinors easily.

Rotation About the Oz Axis

Let us consider a system of axes $Oxyz$ in which an electron is at rest and which is such that the positive direction of Oz coincides with the direction of the magnetic induction $\boldsymbol{B}$. Pauli's equation is then reduced to

$$\left(\frac{e\hbar}{2mc}\boldsymbol{\sigma}\cdot\boldsymbol{B}\right)\psi = E\psi, \tag{4.2.36}$$

where the charge of the election is equal to $-e$ and where a stationary state of energy E has been considered. Let us write B for the component of the field $\boldsymbol{B}$ along the Oz axis, and let us introduce the Bohr magneton $\mu = e\hbar/2mc$. Equation (4.2.36) then reduces to

$$\mu\sigma_3 B \begin{bmatrix} \psi_1 \\ \psi_2 \end{bmatrix} = \mu B \begin{bmatrix} 1 & 0 \\ 0 & -1 \end{bmatrix} \begin{bmatrix} \psi_1 \\ \psi_2 \end{bmatrix} = E \begin{bmatrix} \psi_1 \\ \psi_2 \end{bmatrix}, \tag{4.2.37}$$

whence the two equations for the components of the wave function

$$\mu B\psi_1 = E\psi_1, \qquad -\mu B\psi_2 = E\psi_2. \tag{4.2.38}$$

This system only has a solution for $E = \pm\mu B$. Since $|\psi_1|^2 + |\psi_2|^2 = 1$ we have

$$\begin{aligned} \psi_1 &= e^{i\alpha}, & \psi_2 &= 0, & \text{for} \quad E &= \mu B, \\ \psi_1 &= 0, & \psi_2 &= e^{i\beta}, & \text{for} \quad E &= -\mu B, \end{aligned} \tag{4.2.39}$$

where α and β are arbitrary arguments. For the present let us consider a second system of axes $Ox'y'z'$ such that the axis Oz' makes an angle θ with the field $\boldsymbol{B}$. Let us write B_x, B_y, B_z for the components of $\boldsymbol{B}$ in this new system of axes. Let us write ϕ_1 and ϕ_2 for the new components of the wave function ϕ in this reference frame. Equation (4.2.36) then gives us

$$\mu(\sigma_1 B_x + \sigma_2 B_y + \sigma_3 B_z) \begin{bmatrix} \phi_1 \\ \phi_2 \end{bmatrix} = E \begin{bmatrix} \phi_1 \\ \phi_2 \end{bmatrix}. \tag{4.2.40}$$

The Pauli matrices then give for the equations of the components

$$\begin{aligned} \mu[(B_x - iB_y)\phi_2 + B_z\phi_1] &= E\phi_1, \\ \mu[(B_x + iB_y)\phi_2 - B_z\phi_1] &= E\phi_2, \end{aligned} \tag{4.2.41}$$

In order that these two homogeneous equations have nonzero solutions it is necessary that the determinant of the system be zero, whence the values of E are

$$E = \pm\sqrt{B_x{}^2 + B_y{}^2 + B_z{}^2} = \pm\mu B. \tag{4.2.42}$$

Let us write $\gamma = \arctan(B_y/B_z)$. For the positive values of E we obtain

$$\frac{\phi_2}{\phi_1} = \tan\tfrac{1}{2}\theta e^{i\gamma}. \tag{4.2.43}$$

Since $|\phi_1|^2 + |\phi_2|^2 = 1$ we must have

$$|\phi_1|^2 = \cos^2\tfrac{1}{2}\theta, \qquad |\phi_2|^2 = \sin^2\tfrac{1}{2}\theta. \tag{4.2.44}$$

Since the arguments are arbitrary we can set

$$\begin{aligned}\phi_1 &= \cos\tfrac{1}{2}\theta\exp(-i\tfrac{1}{2}(\gamma-\pi/2)),\\ \phi_2 &= i\sin\tfrac{1}{2}\theta\exp(i\tfrac{1}{2}(\gamma-\pi/2)).\end{aligned}\tag{4.2.45}$$

By a calculation of the same type we obtain for the negative values of E

$$\begin{aligned}\phi_1 &= i\sin\tfrac{1}{2}\theta\exp(-i\tfrac{1}{2}(\gamma-\pi/2)),\\ \phi_2 &= \cos\tfrac{1}{2}\theta\exp(i\tfrac{1}{2}(\gamma-\pi/2)).\end{aligned}\tag{4.2.46}$$

Comparing the expressions for the wave function given by (4.2.39) and those given by (4.2.45) and (4.2.46) allows us to write the transformation matrix of the components under a rotation through an angle θ. Since the angles α and β in expressions (4.2.39) are arbitrary, let us take them as equal to zero in order to simplify the matrix. Setting $\varphi = \frac{1}{2}(\gamma - \pi/2))$ we obtain

$$R(0,0,\tfrac{1}{2}\theta) = \begin{bmatrix}\cos\frac{1}{2}\theta e^{-i\varphi} & i\sin\frac{1}{2}\theta e^{-i\varphi}\\ i\sin\frac{1}{2}\theta e^{i\varphi} & \cos\frac{1}{2}\theta e^{i\varphi}\end{bmatrix}.\tag{4.2.47}$$

As is easily verified, the matrix (4.2.47) is a unimodular unitary matrix. It is a particular case of the matrix (1.2.13) defining the rotation properties of a spinor. The components ψ_1 and ψ_2 of the wave function ψ are thus the components of a spinor.

Rotation About an Arbitrary Axis

A general calculation can be carried out by considering the transformation operator, denoted $U_R(\varphi)$, of the wave function of Pauli's equation, that is,

$$\phi = U_R(\varphi)\psi.\tag{4.2.48}$$

As the integral $\int \psi^\dagger\psi dq$ is invariant, it follows that the operator $U_R(\varphi)$ is a unitary operator. In fact, we have

$$\begin{aligned}\int \psi^\dagger\psi dq &= \int \phi^\dagger\phi dq\\ &= \int \psi^\dagger U_R^\dagger(\varphi)U_R(\varphi)\psi dq.\end{aligned}\tag{4.2.49}$$

Consequently,

$$U_R^\dagger(\varphi)U_R(\varphi) = I,\tag{4.2.50}$$

where I is the unit operator. Looking for a unitary operator we prove that the operator $U_R(\varphi)$ is obtained in the classical form

$$U_R(\varphi) = \exp\left[-\frac{i}{\hbar}\varphi(\boldsymbol{L} + \tfrac{1}{2}\hbar\boldsymbol{\sigma})\right],\tag{4.2.51}$$

where $\boldsymbol{L}$ is the vector operator of the orbital angular momentum and $\boldsymbol{\sigma}$ is the Pauli vector operator. Thus we recover the transformation operator of the spinors under a rotation through an angle φ about an arbitrary axis.

4.3 Spinor and Vector Fields

4.3.1 The Transformation of a Vector Field by a Rotation

An electromagnetic field is classically defined by a vector field, electric and magnetic fields, and more compactly by a tensor field. On the other hand, the electromagnetic field can also be considered as a set of quantum particles having a spin equal to one, namely, the photons. We are going to see that intrinsically a vector field possesses spin properties under a rotation of the field.

Let us consider a vector field $\boldsymbol{A}(\boldsymbol{r}) = (A_1, A_2, A_3)$. Let us write R for the rotation operator in the forward sense and R^{-1} for the inverse operator. Let us consider a rotation of the vector field (an 'active rotation'). At the point defined by the position vector $\boldsymbol{r}$ this rotation generates a transformation of $\boldsymbol{A}(\boldsymbol{r})$. Let us write $U_R(\varphi)$ for the operator which transforms the vector $\boldsymbol{A}(\boldsymbol{r})$ under a rotation of the field of such a kind that

$$U_R(\varphi)\boldsymbol{A}(\boldsymbol{r}) = R\boldsymbol{A}(R^{-1}\boldsymbol{r}). \tag{4.3.1}$$

At the point $\boldsymbol{r}$ this operation corresponds to a local rotation of the vector field.

Infinitesimal Rotation of a Vector Field

Let us write $R(\mathbf{d}\varphi)$ for the rotation through an infinitesimal angle $\mathrm{d}\varphi$ about an arbitrary axis Δ. Let $\mathbf{d}\varphi$ be the vector along this axis and with modulus equal to the angle $\mathrm{d}\varphi$. In accordance with (3.1.20) we then have

$$R(\mathbf{d}\varphi)\boldsymbol{r} = \boldsymbol{r} + \mathrm{d}\boldsymbol{r} = \boldsymbol{r} + \mathbf{d}\varphi \times \boldsymbol{r}. \tag{4.3.2}$$

Let us determine the expression for the operator $U_R(\mathbf{d}\varphi)$ for the infinitesimal rotations defined by the vector $\mathbf{d}\varphi$. We have

$$\begin{aligned} U_R(\mathbf{d}\varphi)\boldsymbol{A}(\boldsymbol{r}) &= R\boldsymbol{A}(R^{-1}\boldsymbol{r}) \\ &= \boldsymbol{A}(R^{-1}\boldsymbol{r}) + \mathbf{d}\varphi \times \boldsymbol{A}(R^{-1}\boldsymbol{r}) \\ &= \boldsymbol{A}(\boldsymbol{r} - \mathbf{d}\varphi \times \boldsymbol{r}) + \mathbf{d}\varphi \times \boldsymbol{A}(\boldsymbol{r} - \mathbf{d}\varphi \times \boldsymbol{r}). \end{aligned} \tag{4.3.3}$$

In the first line of (4.3.3) we have neglected the quantities of second order in $\mathbf{d}\varphi$. Introducing the angular momentum operator $\boldsymbol{L} = \boldsymbol{r} \times i\hbar\boldsymbol{\nabla}$ we have

$$\begin{aligned} \boldsymbol{A}(\boldsymbol{r} - \mathbf{d}\varphi \times \boldsymbol{r}) &= \boldsymbol{A}(\boldsymbol{r}) - (\mathbf{d}\varphi \times \boldsymbol{r})\cdot\boldsymbol{\nabla}\boldsymbol{A}(\boldsymbol{r}) \\ &= \boldsymbol{A}(\boldsymbol{r}) - \frac{i}{\hbar}(\mathbf{d}\varphi\cdot\boldsymbol{L})\boldsymbol{A}(\boldsymbol{r}). \end{aligned} \tag{4.3.4}$$

On the other hand, neglecting the second-order terms in $\mathbf{d}\varphi$, for the last term of (4.3.3) we have

$$\mathbf{d}\varphi \times \boldsymbol{A}(\boldsymbol{r} - \mathbf{d}\varphi \times \boldsymbol{r}) = \mathbf{d}\varphi \times \boldsymbol{A}(\boldsymbol{r}). \tag{4.3.5}$$

Let us write the components of the latter vector in more detail. This yields

$$[\mathbf{d}\varphi \times \boldsymbol{A}(\boldsymbol{r})]_i = \varepsilon_{ijk}\mathrm{d}\varphi_j A_k, \tag{4.3.6}$$

where the $\mathrm{d}\varphi_j$ are the components along the Cartesian axes of the vector $\mathbf{d}\varphi$ and ε_{ijk} is the Kronecker symbol. Let us replace the latter by the following matrix elements

$$(S_j)_{ik} = i\hbar\,\varepsilon_{ijk}, \qquad i, j, k = 1, 2, 3. \tag{4.3.7}$$

Thus, for example, we have the following matrix for $j = 1$

$$\begin{aligned} S_1 &= [(S_1)_{ik}] = i\hbar\,\varepsilon_{i1k} \\ &= i\hbar \begin{bmatrix} 0 & 0 & 0 \\ 0 & 0 & -1 \\ 0 & 1 & 0 \end{bmatrix}. \end{aligned} \tag{4.3.8}$$

We obtain the infinitesimal rotation matrix about the axis Ox given by (3.1.4). Similarly we obtain the matrices S_2 and S_3 which correspond to the infinitesimal rotation matrices about the respective axes Oy and Oz given by formulas (3.1.5). Let us write $\boldsymbol{S}$ for the vector with components which are the matrices S_j, that is, $\boldsymbol{S} = [S_1, S_2, S_3]$. Expressing the components (4.3.6) as a function of the matrix elements (4.3.7) we obtain for the vector expression,

$$\mathbf{d}\varphi \times \boldsymbol{A}(\boldsymbol{r}) = -\frac{i}{\hbar}\,(\mathbf{d}\varphi \cdot \boldsymbol{S})\boldsymbol{A}(\boldsymbol{r}). \tag{4.3.9}$$

Finally, taking account of (4.3.4) and (4.3.9) expression (4.3.3) becomes

$$U_R(\mathbf{d}\varphi \boldsymbol{A}(\boldsymbol{r}) = \left[\mathbb{1} - \frac{i}{\hbar}\,\mathbf{d}\varphi \cdot (\boldsymbol{L} + \boldsymbol{S})\right] \boldsymbol{A}(\boldsymbol{r}), \tag{4.3.10}$$

where $\mathbb{1}$ is the unit matrix of order 3. The expression for the infinitesimal rotation operator of a vector field is therefore

$$U_R(\mathbf{d}\varphi) = \mathbb{1} - \frac{i}{\hbar}\,\mathbf{d}\varphi \cdot (\boldsymbol{L} + \boldsymbol{S}). \tag{4.3.11}$$

For rotation through a finite angle φ we can write that the latter can be divided into N elementary rotations such that $\mathrm{d}\varphi = \varphi/N$. We shall then obtain the operator $U_R(\varphi)$ by N successive rotations. Thus we have

$$\begin{aligned} U_R(\varphi) &= \lim_{N\to\infty}\left[\mathbb{1} - \frac{i}{\hbar}\,\mathrm{d}\boldsymbol{\varphi}\cdot(\boldsymbol{L}+\boldsymbol{S})\right]^N \\ &= \exp\left[-\frac{i}{\hbar}\,\vec{\varphi}\cdot(\boldsymbol{L}+\boldsymbol{S})\right], \end{aligned} \tag{4.3.12}$$

where $\vec{\varphi}$ is the rotation vector, the vector being along the axis of the rotation and with modulus equal to the angle φ.

The Spin of the Vector Field

The components of the vector $\boldsymbol{S}$ satisfy the same commutation relations as those of the vector $\boldsymbol{L}$, since the components are represented by identical matrices. We thus have for the operators and their matrices

$$[S_i, S_j] = i\hbar\,\varepsilon_{ijk}S_k. \tag{4.3.13}$$

The latter commutation relations characterize the components of a quantum angular momentum operator. As the vector $\boldsymbol{L}$ represents an orbital angular momentum, by analogy with the total angular momentum $\boldsymbol{J} = \boldsymbol{L}+\boldsymbol{S}$ of an electron we can interpret the vector $\boldsymbol{S}$ which appears in (4.3.11) as an intrinsic angular momentum, or the spin of the vector field. We are going to see that this interpretation is strengthened by the study of a spinor field which follows.

Before doing so let us calculate the matrix $\boldsymbol{S}^2$ defined as the sum of the squares of its components, that is,

$$\boldsymbol{S}^2 = S_1^2 + S_2^2 + S_3^2 = 2\hbar^2\mathbb{1}. \tag{4.3.14}$$

As the general expression for the value of the square of an angular momentum has the form $s(s+1)\hbar^2$, expressions (4.3.14) can be written in this form, that is,

$$\boldsymbol{S}^2 = 1(1+1)\hbar^2\mathbb{1}, \tag{4.3.15}$$

and the value of s is equal to one. The vector fields thus have a spin equal to unity, and in quantum terms can be described by particles with spin equal to one.

4.3.2 The Rotation of a Spinor Field

Let us consider a spinor field formed of spinors with three components, $\chi(\boldsymbol{r}) = (\psi_1, \psi_2, \psi_3)$. These spinors allow us to form a matrix representation $D^{(1)}$ of the group $\boldsymbol{SU}(2)$ of which the matrices $M^{(1)}(U)$ are given in the general form by (2.2.19).

Under a rotation of this field in the space of the spinors, the latter are transformed according to

$$\chi'(\boldsymbol{r}) = -M^{(1)}(U)\chi(R^{-1}\boldsymbol{r}). \tag{4.3.16}$$

Let us see the form of the infinitesimal rotation operators for this spinor field. For example, let us consider an infinitesimal rotation about the axis Oz through an angle γ. The operator corresponding to the matrix $M^{(1)}(U)$ can be expanded, keeping only the terms of first order in γ, that is,

$$\chi'(\boldsymbol{r}) = (\mathbb{1} + \gamma {A_z}^{(1)})\chi + \gamma \left(y\frac{\partial}{\partial x} - x\frac{\partial}{\partial y} \right) \chi, \tag{4.3.17}$$

where ${A_z}^{(1)}$ is the infinitesimal operator of the representation $\Gamma^{(1)}$ of the group of rotations $\boldsymbol{SO}(3)$. As a consequence, the infinitesimal operator A_z for a spinor field has the form

$$\begin{aligned} A_z &= {A_z}^{(1)} + \left(y\frac{\partial}{\partial x} - x\frac{\partial}{\partial y} \right) \\ &= {A_z}^{(1)} + L_z. \end{aligned} \tag{4.3.18}$$

Similarly we obtain the expressions for the operators A_z and A_y

$$A_x = {A_x}^{(1)} + L_x, \qquad A_y = {A_y}^{(1)} + L_y. \tag{4.3.19}$$

Calling the infinitesimal angles of rotation about the axes Ox, Oy, Oz respectively α, β, γ, we obtain for the rotation operator of a spinor field

$$\begin{aligned} U(\alpha, \beta, \gamma) &= \mathbb{1} + [\alpha({A_x}^{(1)} + L_x) \\ &\qquad + \beta({A_y}^{(1)} + L_y) + \gamma({A_x}^{(1)} + L_z)]. \end{aligned} \tag{4.3.20}$$

In addition, the orbital angular momentum operator $\boldsymbol{L}$ with components L_x, L_y, L_z, the infinitesimal operator $U(\alpha, \beta, \gamma)$ involves the vector operator denoted $\boldsymbol{A}$ having as its components the operators A_x, A_y, A_z. Let us also write $\mathbf{d}\varphi$ for the infinitesimal rotation vector having as the moduli of its components the quantities α, β, γ. Formula (4.3.20) is then written

$$U(\alpha, \beta, \gamma) = \mathbb{1} + \mathbf{d}\varphi\cdot(\boldsymbol{L} + \boldsymbol{A}). \tag{4.3.21}$$

The operators A_x, A_y, A_z are equivalent to the infinitesimal operators of the various three-dimensional irreducible representations of the group $\boldsymbol{SO}(3)$, as we have seen in the course of Chapter 3. In particular, they are equivalent to the operators S_i of which the matrix elements are defined by (4.3.7). The formulas (4.3.11) and (4.3.21) are thus identical to within a change of basis.

Finally, the identification of the vector $\boldsymbol{S}$ as a vector operator for the spin is verified because the operator $\boldsymbol{A}$ is an operator acting specifically in the space of the spinors. This is linked essentially with the property that the three-dimensional vectors and the three-component spinors represent one and the same representation of the rotation group.

4.4 Solved Problems

Exercise **4.1**

1. Prove that the Pauli matrices σ_j, $j = 1, 2, 3$, satisfy the relation

$$\sigma_j \sigma_k = i \sum_{m=1}^{3} \varepsilon_{jkm} \sigma_m + \delta_{jk} \mathbb{1}.$$

2. Prove that $\sigma_1 \sigma_2 \sigma_3 = i\mathbb{1}$, where $\mathbb{1}$ is the unit matrix of order 2.

3. Prove that every matrix M of order 2 can be put into the form

$$M = a_0 \mathbb{1} + \boldsymbol{a} \cdot \boldsymbol{\sigma},$$

where $\boldsymbol{a} = (a_1, a_2, a_3)$ and $\boldsymbol{\sigma} = (\sigma_1, \sigma_2, \sigma_3)$.

Solution **4.1**

1. Write the Pauli matrices as

$$\sigma_x = \sigma_1, \qquad \sigma_y = \sigma_2, \qquad \sigma_z = \sigma_3.$$

On the one hand these matrices satisfy the relations

$$\sigma_1{}^2 = \sigma_2{}^2 = \sigma_3{}^3 = \mathbb{1}.$$

On the other hand, the Pauli matrices anticommute. For example, we have

$$\sigma_1 \sigma_2 = -\sigma_2 \sigma_1 = i\sigma_3 = i \sum_{m=1}^{3} \varepsilon_{12m} \sigma_m,$$

as well as the analogous equations obtained by cyclic permutation of the indices 1, 2, and 3. The preceding relations can then be written in the condensed form

$$\sigma_j \sigma_k = i \sum_{m=1}^{3} \varepsilon_{jkm} \sigma_m + \delta_{jk} \mathbb{1}, \qquad \text{with} \quad j, k = 1, 2, 3.$$

2. The relations

$$\sigma_1 \sigma_2 = i\sigma_3 \qquad \text{and} \qquad \sigma_3{}^2 = \mathbb{1}$$

give us

$$\sigma_1 \sigma_2 \sigma_3 = i\sigma_3{}^2 = i\mathbb{1}.$$

3. Consider the matrix of order 2

$$M = \begin{bmatrix} a_{11} & a_{12} \\ a_{21} & a_{22} \end{bmatrix}.$$

We easily verify that this matrix can be put into the form

$$M = \tfrac{1}{2}(a_{11}+a_{22})\mathbb{1}+\tfrac{1}{2}(a_{12}+a_{21})\sigma_x+\tfrac{1}{2}i(a_{12}-a_{21})\sigma_y+\tfrac{1}{2}(a_{11}-a_{22})\sigma_z.$$

Thus we can write

$$\begin{aligned} M &= a_0\mathbb{1} + a_x\sigma_x + a_y\sigma_y + a_z\sigma_z \\ &= a_1\mathbb{1} + \boldsymbol{a}\cdot\boldsymbol{\sigma}. \end{aligned}$$

EXERCISE **4.2**
Let us write

$$|+\rangle = (1,0) \qquad \text{and} \qquad |-\rangle = (0,1)$$

for the orthonormal spinors which can serve as a basis of the space of spinors.

1. Show that $|+\rangle$ and $|-\rangle$ are eigenvectors of the Pauli matrix σ_z and calculate its eigenvectors.

2. Calculate the eigenvalues and the normed eigenspinors of σ_x and σ_y in the basis $\{|+\rangle, |-\rangle\}$.

SOLUTION **4.2**

1. Applying the Pauli matrix to the spinor $|+\rangle$ we obtain

$$\sigma_z |+\rangle = \begin{bmatrix} 1 & 0 \\ 0 & -1 \end{bmatrix}\begin{bmatrix} 1 \\ 0 \end{bmatrix} = \begin{bmatrix} 1 \\ 0 \end{bmatrix} = |+\rangle .$$

The eigenvalue which corresponds to the eigenvector $|+\rangle$ is equal to 1. Similarly we have

$$\sigma_z |-\rangle = \begin{bmatrix} 1 & 0 \\ 0 & -1 \end{bmatrix}\begin{bmatrix} 0 \\ 1 \end{bmatrix} = -\begin{bmatrix} 0 \\ 1 \end{bmatrix} = -|-\rangle .$$

The corresponding eigenvalue of the eigenvector $|-\rangle$ is equal to -1. These are the eigenvalues which appear on the principal diagonal of the matrix σ_z.

2. The eigenvalues of the matrices σ_x and σ_y are calculated from the characteristic equations

$$\begin{vmatrix} -\lambda & 1 \\ 1 & -\lambda \end{vmatrix} = \begin{vmatrix} -\lambda & -i \\ i & -\lambda \end{vmatrix} = 0,$$

that is to say, $\lambda^2 = 1$, whence the eigenvalues of σ_x and σ_y are $\lambda = \pm 1$. The eigenspinors of the matrices σ_x and σ_y corresponding to these eigenvalues are respectively written $|\pm\rangle_x$ and $|\pm\rangle_y$. They are sought in the form of a linear combination of the spinors $|+\rangle$ and $|-\rangle$, thus, for example,

$$|+\rangle_x = a\,|+\rangle + b\,|-\rangle\,.$$

For the eigenvalue $\lambda = 1$ we must have

$$\sigma_x(a\,|+\rangle + b\,|-\rangle) = a\,|+\rangle + b\,|-\rangle\,.$$

Expanding the latter expression in matrix form we obtain $a = b$, whence the normed expression for the eigenvector

$$|+\rangle_x = \frac{1}{\sqrt{2}}(|+\rangle + |-\rangle).$$

Similarly, for $\lambda = -1$ we obtain $a = -b$, whence the eigenvector

$$|-\rangle_x = \frac{1}{\sqrt{2}}(|+\rangle - |-\rangle).$$

The eigenvectors of σ_y are obtained similarly, whence

$$|\pm\rangle_y = \frac{1}{\sqrt{2}}(|+\rangle \pm i\,|-\rangle).$$

Exercise **4.3**

Consider a unitary vector $\boldsymbol{u}$ of which the components in spherical coordinates θ, φ are

$$u_x = \sin\theta\cos\varphi, \qquad u_y = \sin\theta\sin\varphi, \qquad u_z = \cos\theta.$$

Let S_x, S_y, S_z be the operators the matrices of which are the Pauli matrices in the basis $\{|+\rangle\,, |-\rangle\}$.

1. Write S_u for the operator

$$S_u = \boldsymbol{u}\cdot\boldsymbol{S} = u_x S_x + u_y S_y + u_z S_z.$$

Calculate the matrix σ_u of the operator S_u in the spinor basis $\{|+\rangle\,, |-\rangle\}$.

2. Calculate the eigenvalues and eigenvectors of the matrix σ_u.

SOLUTION **4.3**

1. The expression for the Pauli matrices and the components of $\boldsymbol{u}$ give us

$$\sigma_u = \sin\theta\cos\varphi\begin{bmatrix} 0 & 1 \\ 1 & 0 \end{bmatrix} + \sin\theta\sin\varphi\begin{bmatrix} 0 & -i \\ i & 0 \end{bmatrix} + \cos\theta\begin{bmatrix} 1 & 0 \\ 0 & -1 \end{bmatrix},$$

whence

$$\sigma_u = \begin{bmatrix} \cos\theta & \sin\theta e^{-i\varphi} \\ \sin\theta e^{i\varphi} & -\cos\theta \end{bmatrix}.$$

2. The eigenvalues obtained from the characteristic equation are those of the Pauli matrices, $\lambda = \pm 1$. The eigenvectors are sought in the form of a linear combination of the vectors $|+\rangle$ and $|-\rangle$, that is,

$$|\pm\rangle_u = a\,|+\rangle + b\,|-\rangle\,.$$

For $\lambda = 1$ we obtain

$$\frac{a}{b} = \frac{1+\cos\theta}{\sin\theta e^{i\varphi}} = \frac{\cos\frac{1}{2}\theta}{\sin\frac{1}{2}\theta e^{i\varphi}}.$$

Since the wave functions are defined to within a phase factor we can take

$$|+\rangle_u = \cos\tfrac{1}{2}\theta\exp(-i\tfrac{1}{2}\varphi)\,|+\rangle + \sin\tfrac{1}{2}\theta\exp(i\tfrac{1}{2}\varphi)\,|-\rangle\,.$$

Similarly, for $\lambda = 1$, we obtain

$$|-\rangle_u = -\sin\tfrac{1}{2}\theta\exp(-i\tfrac{1}{2}\varphi)\,|+\rangle + \cos\tfrac{1}{2}\theta\exp(i\tfrac{1}{2}\varphi)\,|-\rangle\,.$$

EXERCISE **4.4**
Prove the formula

$$(\boldsymbol{\sigma}\cdot\boldsymbol{A})(\boldsymbol{\sigma}\cdot\boldsymbol{B}) = \boldsymbol{A}\cdot\boldsymbol{B} + i\boldsymbol{\sigma}\cdot(\boldsymbol{A}\times\boldsymbol{B}).$$

SOLUTION **4.4**
The Pauli matrices satisfy the following relation calculated in Exercise 4.1

$$\sigma_j\sigma_k = i\sum_{m=1}^{3}\varepsilon_{jkm}\sigma_m + \delta_{jk}\mathbb{1}.$$

As the components A_i and B_j of the vectors $\boldsymbol{A}$ and $\boldsymbol{B}$ are numbers, we can write

$$\begin{aligned}(\boldsymbol{\sigma}\cdot\boldsymbol{B})(\boldsymbol{\sigma}\cdot\boldsymbol{B}) &= \left(\sum_i \sigma_i A_i\right)\left(\sum_i \sigma_j B_j\right)\\ &= \sum_{i,j} A_i B_j \sigma_i \sigma_j .\end{aligned}$$

Taking into account the expression for the products of the Pauli matrices we have

$$\sum_{i,j} A_i B_j \sigma_i \sigma_j = \sum_{i,j} i A_i B_j \left(\sum_k \varepsilon_{ijk}\sigma_k\right) + \sum_{i,j} A_i B_j \delta_{ij} \mathbb{1}.$$

The last term of the preceding expression is equal to the scalar product of the vectors $\boldsymbol{A}$ and $\boldsymbol{B}$. In fact,

$$\sum_{i,j} A_i B_j \delta_{ij} = \sum_i A_i B_i = \boldsymbol{A}\cdot\boldsymbol{B}.$$

On the other hand, the component $(\boldsymbol{A}\times\boldsymbol{B})_k$ of the vector product $\boldsymbol{A}\times\boldsymbol{B}$ can be written in the form

$$(\boldsymbol{A}\times\boldsymbol{B})_k = \sum_{i,j} A_i B_j \varepsilon_{ijk},$$

as can be easily verified. We can therefore write

$$\begin{aligned}\sum_{i,j} i A_i B_j \left(\sum_k \varepsilon_{ijk}\sigma_k\right) &= i\sum_k \sigma_k \left(\sum_{i,j} A_i B_j \varepsilon_{ijk}\right)\\ &= i\sum_k \sigma_k (\boldsymbol{A}\times\boldsymbol{B})_k\\ &= i\boldsymbol{\sigma}\cdot(\boldsymbol{A}\times\boldsymbol{B}).\end{aligned}$$

Collecting together the scalar product and the preceding mixed product, finally we obtain

$$\begin{aligned}(\boldsymbol{\sigma}\cdot\boldsymbol{A})(\boldsymbol{\sigma}\cdot\boldsymbol{B}) &= \left(\sum_i \sigma_i A_i\right)\left(\sum_i \sigma_j B_j\right)\\ &= \boldsymbol{A}\cdot\boldsymbol{B}\mathbb{1} + i\boldsymbol{\sigma}\cdot(\boldsymbol{A}\times\boldsymbol{B}).\end{aligned}$$

The unit matrix $\mathbb{1}$ of order 2 does not generally appear in the preceding expression, but it is clearly implicit.

Exercise **4.5**
The two-component Pauli spinors ψ and η satisfy equations (4.2.25). Show that these spinors also satisfy the classical Schrödinger equation.

Solution **4.5**
Equations (4.2.25) have as their expression

$$\boldsymbol{\sigma}\cdot\boldsymbol{P}\psi + 2im\eta = 0, \qquad \boldsymbol{\sigma}\cdot\boldsymbol{P}\eta - iE\psi = 0.$$

The first equation gives us

$$\eta = -\frac{\boldsymbol{\sigma}\cdot\boldsymbol{P}}{2im}\psi.$$

Substituting this expression for η into the second equation yields

$$\left[E - \frac{(\boldsymbol{\sigma}\cdot\boldsymbol{P})(\boldsymbol{\sigma}\cdot\boldsymbol{P})}{2m}\right]\psi = 0.$$

Using the relation obtained in Exercise 4.4 with $\boldsymbol{A} = \boldsymbol{B} = \boldsymbol{P}$ yields

$$(\boldsymbol{\sigma}\cdot\boldsymbol{P})(\boldsymbol{\sigma}\cdot\boldsymbol{P}) = \boldsymbol{P}\cdot\boldsymbol{P} + i\boldsymbol{\sigma}\cdot(\boldsymbol{P}\times\boldsymbol{P}).$$

As the term $(\boldsymbol{P}\times\boldsymbol{P})$ is clearly zero the preceding equation becomes

$$\left(\frac{\boldsymbol{P}^2}{2m} - E\right)\psi = 0.$$

The spinor ψ can also be eliminated from equations (4.2.25). We have

$$\psi = -\frac{i\boldsymbol{\sigma}\cdot\boldsymbol{P}}{E}\eta.$$

Substituting ψ in the first equation yields

$$[(\boldsymbol{\sigma}\cdot\boldsymbol{P})(\boldsymbol{\sigma}\cdot\boldsymbol{P}) - 2mE]\eta = 0,$$

that is

$$\left(\frac{\boldsymbol{P}^2}{2m} - E\right)\eta = 0.$$

Putting the spinors in their component form

$$\psi = \begin{bmatrix} \psi_1 \\ \psi_2 \end{bmatrix}, \qquad \eta = \begin{bmatrix} \eta_1 \\ \eta_2 \end{bmatrix},$$

we see that each component satisfies the classical Schrödinger equation

$$\left(\frac{\boldsymbol{P}^2}{2m} - E\right)\psi_i = 0, \qquad \left(\frac{\boldsymbol{P}^2}{2m} - E\right)\eta_i = 0, \quad i = 1, 2.$$

Part II

Spinors in Four-Dimensional Space

5

The Lorentz Group

5.1 The Generalized Lorentz Group

5.1.1 Rotations and Reflections

The space–time of Relativity Theory is the space $\mathbb{R}^4$ where each event is referred to the coordinates x^0, x^1, x^2, x^3. The time t is related to x^0 by $x^0 = ct$, where c is the speed of light and the spacial coordinates x^1, x^2, x^3 define the position vector $\boldsymbol{x}$. The vectors $\boldsymbol{X} = (x^0, x^1, x^2, x^3)$ are called *four-vectors*.

According to the theory of Special Relativity the spacial coordinates and the time in two frames which are displaced from each other by a uniform motion are related by a linear transformation which we shall call a PROPER LORENTZ TRANSFORMATION. The latter has a well defined form, classically used in physics and it leaves invariant the quadratic form

$$\varphi(\boldsymbol{X}) = -(x^0)^2 + (x^1)^2 + (x^2)^2 + (x^3)^2. \tag{5.1.1}$$

We call every linear transformation which leaves the quadratic form (5.1.1) a GENERAL LORENTZ TRANSFORMATION. Let us write this general transformation $\boldsymbol{X} \to \boldsymbol{X}' = \Lambda \boldsymbol{X}$. We have

$$\begin{aligned}\varphi(\Lambda \boldsymbol{X}) &= -(x'^0)^2 + (x'^1)^2 + (x'^2)^2 + (x'^3)^2 \\ &= \varphi(\boldsymbol{X}).\end{aligned} \tag{5.1.2}$$

Let us write X' for the row matrix $[x^0 \; x^1 \; x^2 \; x^3]$ and X for the corresponding column matrix which is the transpose of X'. Let us write G for the matrix

$$G = \begin{bmatrix} -1 & 0 & 0 & 0 \\ 0 & 1 & 0 & 0 \\ 0 & 0 & 1 & 0 \\ 0 & 0 & 0 & 1 \end{bmatrix}. \tag{5.1.3}$$

The quadratic form (5.1.1) can then be written

$$\varphi(\boldsymbol{X}) = X'GX. \tag{5.1.4}$$

Similarly, the form $\varphi(\Lambda\boldsymbol{X})$ may be written

$$\varphi(\Lambda\boldsymbol{X}) = (\Lambda X)'G(\Lambda X). \tag{5.1.5}$$

Taking account of (5.1.4) and (5.1.5), equality (5.1.2) then gives

$$(\Lambda X)'G(\Lambda X) = X'\Lambda'G\Lambda X = X'\Lambda X. \tag{5.1.6}$$

Since the matrix X is arbitrary we have the following relation which is characterized by a general Lorentz transformation

$$\Lambda'G\Lambda = G. \tag{5.1.7}$$

The set of matrices Λ which satisfy the latter relation forms the GENERALIZED LORENTZ GROUP. This group is a real Lie group, and is denoted $\boldsymbol{O}(3,1)$. We shall see that this group can be decomposed into several subsets, one of which contains the proper Lorentz transformations.

To do so let us calculate the determinant of the matrices Λ. As the determinant of a product of matrices is equal to the product of their determinants, relation (5.1.7) gives us $(\det\Lambda)^2 = 1$, whence

$$\det\Lambda = \pm 1. \tag{5.1.8}$$

We can therefore decompose the group $\boldsymbol{O}(3,1)$ into two subsets characterised by $\det\Lambda = 1$ or $\det\Lambda = -1$.

The *Lorentz rotations* are, by definition, the general Lorentz transformations with determinant equal to 1. These rotations constitute a subgroup of $\boldsymbol{O}(3,1)$ which is denoted $\boldsymbol{L}_+$. It is a normal subgroup and the quotient group $\boldsymbol{O}(3,1)/\boldsymbol{L}_+$ has two elements—the unit matrix and the matrix G.

The Lorentz transformations with determinant -1 clearly does not constitute a subgroup since they do not include the unit matrix. These transformations are called *Lorentz reflections*.

REMARK **5.1.** The groups $\boldsymbol{O}(p,q)$ are orthogonal groups which leave invariant the linear form

$$\varphi(\boldsymbol{X}) = {x_1}^2 + \cdots + {x_p}^2 - {x_{p+1}}^2 - \cdots - {x_{p+q}}^2. \tag{5.1.9}$$

The notation $\boldsymbol{O}(3,1)$ is the mathematical notation for the generalized Lorentz group where the spacial variables are numbered x_1, x_2, x_3 and the time variable is numbered x_4. Here we have used another order which is more usual in physics, putting the time as the first variable and then followed by the space variables. ■

5.1.2 *Orthochronous and Anti-Orthochronous Transformations*

The matrix product $\Lambda' G \Lambda$ which appears in relation (5.1.7) gives for the expression of the first matrix element of this product

$$\Lambda_{00}{}^2 - (\Lambda_{10}{}^2 + \Lambda_{20}{}^2 + \Lambda_{30}{}^2) = 1. \tag{5.1.10}$$

Consequently we have the relation $\Lambda_{00}{}^2 \geqslant 1$, and as a result,

$$\Lambda_{00} \geqslant 1 \quad \text{and} \quad \Lambda_{00} \leqslant -1. \tag{5.1.11}$$

We call ORTHOCHRONOUS LORENTZ TRANSFORMATIONS those for which $\Lambda_{00} \geqslant 1$ and ANTI-ORTHOCHRONOUS LORENTZ TRANSFORMATIONS those for which $\Lambda_{00} \leqslant -1$. The former preserve the sign of x^0, that is to say, the time direction, and the latter reverse it.

The orthochronous transformations form a subgroup of $\boldsymbol{O}(3,1)$ called the ORTHOCHRONOUS LORENTZ GROUP, denoted $\boldsymbol{L}^{\uparrow}$, where the direction of the arrow indicates the direction of the flow of physical time. The group $\boldsymbol{L}^{\uparrow}$ is distinguished and the quotient group $\boldsymbol{O}(3,1)/\boldsymbol{L}^{\uparrow}$ has two elements, the matrices I and $-I$.

The anti-orthochronous transformations form a set, denoted $\boldsymbol{L}^{\downarrow}$, which does not form a subgroup since it does not contain the unit matrix.

5.1.3 *Sheets of the Generalized Lorentz Group*

Combining the characteristics $\det \Lambda = \pm 1$ and $\Lambda_{00} \leqslant -1$ or $\Lambda_{00} \geqslant 1$, we obtain four subsets of the group $\boldsymbol{O}(3,1)$:

— the orthochronous rotations, the set of which is denoted $\boldsymbol{L}^{\uparrow}{}_{+}$ ($\det \Lambda = 1$, $\Lambda_{00} \geqslant 1$);

— the orthochronous reflections, denoted $\boldsymbol{L}^{\uparrow}{}_{-}$ ($\det \Lambda = -1$, $\Lambda_{00} \geqslant 1$);

— the anti-orthochronous rotations, denoted $\boldsymbol{L}^{\downarrow}{}_{+}$ ($\det \Lambda = 1$, $\Lambda_{00} \leqslant -1$);

— the anti-orthochronous reflections, denoted $\boldsymbol{L}^{\downarrow}{}_{-}$ ($\det \Lambda = -1$, $\Lambda_{00} \leqslant -1$).

The four sets are mutually disjoint. Clearly we have

$$\boldsymbol{L}_{+} = \boldsymbol{L}^{\uparrow}{}_{+} \cup \boldsymbol{L}^{\downarrow}{}_{+}, \qquad \boldsymbol{L}^{\uparrow} = \boldsymbol{L}^{\uparrow}{}_{+} \cup \boldsymbol{L}^{\uparrow}{}_{-}. \tag{5.1.12}$$

We prove that we can pass continuously from one element to any other element of the same set, whence the name of *sheet* given to each of them. In contrast, we cannot pass continuous from an element of one sheet to an element of another sheet.

The Proper Lorentz Group

Amongst the four sheets of $\boldsymbol{O}(3,1)$ only the set $\boldsymbol{L}_{+}^{\uparrow}$ is a subgroup. Its determinant is equal to unity and thus is a special orthogonal group. Its standard group notation is $\boldsymbol{SO}(3,1)^{\uparrow}$. Let us prove that it actually is a group.

In fact, let Λ and Λ' be two matrices belonging to $\boldsymbol{SO}(3,1)^{\uparrow}$. These matrices are such that $\Lambda_{00} \geqslant 1$ and $\Lambda'_{00} \geqslant 1$. It is then necessary to prove that the element $(\Lambda\Lambda')_{00}$ of the product matrix $\Lambda\Lambda'$ also satisfies $(\Lambda\Lambda')_{00} \geqslant 1$. This element is equal to

$$(\Lambda\Lambda')_{00} = \Lambda_{00}\Lambda'_{00} + \Lambda_{01}\Lambda'_{01} + \Lambda_{01}\Lambda'_{02} + \Lambda_{03}\Lambda'_{03}. \tag{5.1.13}$$

On the other hand, by taking the transpose of the matrices equality (5.1.7) can be written

$$\Lambda G\Lambda' = G' = G, \tag{5.1.14}$$

which gives us

$${\Lambda_{00}}^2 - ({\Lambda_{01}}^2 + {\Lambda_{02}}^2 + {\Lambda_{03}}^2) = 1. \tag{5.1.15}$$

Applying Schwarz' inequality and using relations (5.1.10) and (5.1.15) yields

$$\begin{aligned} &\Lambda_{01}\Lambda'_{10} + \Lambda_{02}\Lambda'_{20} + \Lambda_{03}\Lambda'_{30} \\ &\qquad \leqslant ({\Lambda_{01}}^2 + {\Lambda_{02}}^2 + {\Lambda_{03}}^2)({\Lambda'_{10}}^2 + {\Lambda'_{02}}^2 + {\Lambda'_{30}}^2) \\ &\qquad < {\Lambda_{00}}^2{\Lambda'_{00}}^2. \end{aligned} \tag{5.1.16}$$

Since $\Lambda_{00} \geqslant 1$ and $\Lambda'_{00} \geqslant 1$ we deduce from (5.1.13) that $(\Lambda\Lambda')_{00} \geqslant 0$. As the matrix $\Lambda\Lambda'$ belongs to $\boldsymbol{SO}(3,1)$ we therefore finally deduce

$$(\Lambda\Lambda')_{00} \geqslant 1. \tag{5.1.17}$$

As a result the sheet $\boldsymbol{SO}(3,1)^{\uparrow}$ does form a group. It is the group of orthochronous rotations. As two arbitrary matrices of this group can be transformed into each other continuously, and since the unit matrix clearly belongs to the group, the latter contains all the proper Lorentz transformations. The group $\boldsymbol{SO}(3,1)^{\uparrow}$ is called the PROPER LORENTZ GROUP, or more briefly the LORENTZ GROUP.

The Construction of $\boldsymbol{O}(3,1)$ *from* $\boldsymbol{SO}(3,1)^{\uparrow}$

Let us introduce the three generalized Lorentz matrices

$$G = \begin{bmatrix} -1 & 0 & 0 & 0 \\ 0 & 1 & 0 & 0 \\ 0 & 0 & 1 & 0 \\ 0 & 0 & 0 & 1 \end{bmatrix}, \quad H = \begin{bmatrix} 1 & 0 & 0 & 0 \\ 0 & -1 & 0 & 0 \\ 0 & 0 & -1 & 0 \\ 0 & 0 & 0 & -1 \end{bmatrix}, \quad -I. \qquad (5.1.18)$$

The matrix G reverses the time. The matrix H generates a spacial inversion. The matrix $-I$ is a total inversion. We can then form the following sets of conjugates from $\boldsymbol{SO}(3,1)^{\uparrow}$

$$G\boldsymbol{SO}(3,1)^{\uparrow} = \boldsymbol{L}^{\downarrow}{}_{-}, \qquad H\boldsymbol{SO}(3,1)^{\uparrow} = \boldsymbol{L}^{\uparrow}{}_{-}, \qquad -I\boldsymbol{SO}(3,1)^{\uparrow} = L^{\uparrow}{}_{+}.$$

The generalized Lorentz group $\boldsymbol{O}(3,1)$ is thus the union of the four sets

$$\boldsymbol{O}(3,1) = I\boldsymbol{SO}(3,1)^{\uparrow} \cup G\boldsymbol{SO}(3,1)^{\uparrow} \cup H\boldsymbol{SO}(3,1)^{\uparrow} \cup -I\boldsymbol{SO}(3,1)^{\uparrow}. \qquad (5.1.19)$$

5.2 The Four-Dimensional Rotation Group

We shall see that the problem of the search for irreducible representations of the proper Lorentz group can be reduced to the analogous problem of the four-dimensional rotation group.

5.2.1 Four-Dimensional Orthogonal Transformations

The coordinates of the space in which the orthogonal transformations act are also denoted by x^0, x^1, x^2, x^3, and they are real numbers. For these transformations the invariant is the following positive definite quadratic form

$$\varphi(\boldsymbol{X}) = (x^0)^2 + (x^1)^2 + (x^2)^2 + (x^3)^2 = x_i x^i, \qquad (5.2.1)$$

with $i = 0, 1, 2, 3$. The notation x_i is the covariant component of $\boldsymbol{X}$ and x^i is its contravariant component [10], and the Einstein summation convention is then used.

Let us consider an orthogonal transformation $A^i{}_j$ which transforms a point with coordinates x^j into a new point x'^j. We have

$$x'^i = A^i{}_j x^j. \qquad (5.2.2)$$

As an orthogonal transformation preserves the norm of the vector, we have

$$\varphi(A\boldsymbol{X}) = x'_i x'^i = \varphi(\boldsymbol{X}). \qquad (5.2.3)$$

Let us write the transformation of the covariant components in the form

$$x'_j = A_j{}^k x_k. \tag{5.2.4}$$

Taking account of (5.2.4), equality (5.2.3) may then be expanded

$$x'^j x'_j = A^j{}_i x^i x_k = x_i x^i, \tag{5.2.5}$$

whence

$$A^j{}_i A_j{}^k = -\delta_i^k, \qquad i, j, k = 0, 1, 2, 3. \tag{5.2.6}$$

This is the classical relation between the orthogonal matrix elements, and it can be used as the definition of an orthogonal transformation. Let us apply successively two orthogonal transformations $A^j{}_i$ and $B^k{}_j$. We obtain

$$x''^k = B^k{}_j x'^j = B^k{}_j A^j{}_i x^i = C^k{}_i x^i. \tag{5.2.7}$$

The transformation $C^k{}_i$ is also an orthogonal transformation, since we have

$$\begin{aligned} C_i{}^k C^i{}_m &= (B_i{}^j A_k{}^k)(B^i{}_l A^l{}_m) = \delta_i{}^j A_j{}^k A^l{}_m \\ &= A_l{}^k A^l{}_m = \delta_m{}^k. \end{aligned} \tag{5.2.8}$$

Thus the orthogonal transformations $A^k{}_j$ form a group. In fact, the identity transformation is given by $A^k{}_j = \delta_j^k$ and the existence of the inverse elements results from relation (5.2.5), corresponding to the matrix relation $AA^{-1} = I$. This group is called the four-dimensional orthogonal transformation group, and is denoted $\boldsymbol{O}(4)$.

The matrices of the orthogonal groups have a determinant equal to 1 or -1. The rotations correspond to the matrices with determinant $+1$ and form a subgroup of $\boldsymbol{O}(4)$. This subgroup constitutes the *four-dimensional rotation group* denoted $\boldsymbol{SO}(4)$. The set of matrices with determinant equal to -1 does not form a group since it does not contain the unit matrix.

5.2.2 Matrix Representations of the Group $\boldsymbol{SO}(4)$

Independent Parameters

The parameters $A^k{}_j$ of an orthogonal transformation are related to each other by relations (5.2.6), which leaves only six independent parameters for defining an orthogonal transformation.

Let us consider as the six independent parameters the six angles of rotation in the two-dimensional planes of a four-dimensional space. This choice is the generalization of what we studied for three-dimensional rotations when we considered rotations about each of the axes Ox, Oy, OZ. Let us keep the notations previously used for the Lorentz group. We have the following planes

$$\begin{aligned} &(x^0, x^1),\ (x^0, x^2),\ (x^0, x^3) \quad \text{— the spacio–temporal planes,} \\ &(x^1, x^2),\ (x^1, x^3),\ (x^2, x^3) \quad \text{— the spacial planes.} \end{aligned} \tag{5.2.9}$$

The order of indices specifies a sense of orientation in these planes. Let us write the angles of rotation by using the indices appearing in the corresponding planes, that is,

$$\varphi_{01},\ \varphi_{02},\ \varphi_{03};\ \varphi_{12},\ \varphi_{13},\ \varphi_{23}. \tag{5.2.10}$$

Matrices of Two-Dimensional Rotations

As we have seen for the three-dimensional rotations of the group $\boldsymbol{SO}(3)$ we can form the rotation matrices in each of the planes (5.2.9). Thus a rotation through an angle ϕ_{23} in the plane (x^2, x^3) leaves the coordinates x^0 and x^1 invariant and transforms the coordinates x^2 and x^3 into them according to the usual rotation formulas in a plane. The matrix, denoted A_{23}, of this rotation of a position vector in a fixed reference frame may thus be written

$$A_{23} = \begin{bmatrix} 1 & 0 & 0 & 0 \\ 0 & 1 & 0 & 0 \\ 0 & 0 & \cos\varphi_{23} & -\sin\varphi_{23} \\ 0 & 0 & \sin\varphi_{23} & \cos\varphi_{23} \end{bmatrix}. \tag{5.2.11}$$

We obtain analogous matrices for the rotations in each of the planes, which are spacial or spacio–temporal planes. We have the following matrices

$$\begin{aligned} A_{13} &= \begin{bmatrix} 1 & 0 & 0 & 0 \\ 0 & \cos\varphi_{13} & 0 & -\sin\varphi_{13} \\ 0 & 0 & 1 & 0 \\ 0 & \sin\varphi_{13} & 0 & \cos\varphi_{13} \end{bmatrix}, \\ A_{12} &= \begin{bmatrix} 1 & 0 & 0 & 0 \\ 0 & \cos\varphi_{12} & -\sin\varphi_{12} & 0 \\ 0 & \sin\varphi_{12} & \cos\varphi_{12} & 0 \\ 0 & 0 & 0 & 0 \end{bmatrix}, \end{aligned} \tag{5.2.12a}$$

$$A_{01} = \begin{bmatrix} \cos\varphi_{01} & -\sin\varphi_{01} & 0 & 0 \\ \sin\varphi_{01} & \cos\varphi_{01} & 0 & 0 \\ 0 & 0 & 1 & 0 \\ 0 & 0 & 0 & 1 \end{bmatrix},$$

$$A_{02} = \begin{bmatrix} \cos\varphi_{02} & 0 & -\sin\varphi_{02} & 0 \\ 0 & 1 & 0 & 0 \\ \sin\varphi_{02} & 0 & \cos\varphi_{02} & 0 \\ 0 & 0 & 0 & 1 \end{bmatrix}. \tag{5.2.12b}$$

$$A_{03} = \begin{bmatrix} \cos\varphi_{03} & 0 & 0 & -\sin\varphi_{03} \\ 0 & 1 & 0 & 0 \\ 0 & 0 & 1 & 0 \\ \sin\varphi_{03} & 0 & 0 & \cos\varphi_{03} \end{bmatrix}.$$

Let us note that these matrices have been calculated from the components of a vector defined in an orthonormal basis.

5.2.3 Infinitesimal Matrices

The derivative of each matrix element of a matrix A_{ij} with respect to the two-dimensional rotation angle gives us the corresponding matrix element of the infinitesimal matrix X_{ij}. Thus we obtain the six infinitesimal matrices

$$X_{23} = \begin{bmatrix} 0 & 0 & 0 & 0 \\ 0 & 0 & 0 & 0 \\ 0 & 0 & 0 & -1 \\ 0 & 0 & 1 & 0 \end{bmatrix}, \qquad X_{13} = \begin{bmatrix} 0 & 0 & 0 & 0 \\ 0 & 0 & 0 & -1 \\ 0 & 0 & 0 & 0 \\ 0 & 1 & 0 & 0 \end{bmatrix}, \tag{5.2.13}$$

$$X_{12} = \begin{bmatrix} 0 & 0 & 0 & 0 \\ 0 & 0 & -1 & 0 \\ 0 & 1 & 0 & 0 \\ 0 & 0 & 0 & 0 \end{bmatrix},$$

$$X_{01} = \begin{bmatrix} 0 & -1 & 0 & 0 \\ 1 & 0 & 0 & 0 \\ 0 & 0 & 0 & 0 \\ 0 & 0 & 0 & 0 \end{bmatrix}, \qquad X_{02} = \begin{bmatrix} 0 & 0 & -1 & 0 \\ 0 & 0 & 0 & 0 \\ 1 & 0 & 0 & 0 \\ 0 & 0 & 0 & 0 \end{bmatrix},$$
$$X_{03} = \begin{bmatrix} 0 & 0 & 0 & -1 \\ 0 & 0 & 0 & 0 \\ 0 & 0 & 0 & 0 \\ 1 & 0 & 0 & 0 \end{bmatrix}. \tag{5.2.13}$$

Properties of Infinitesimal Matrices

We remark that the matrices (5.2.13) are antisymmetric. On the other hand, reversing the indices corresponds to a change of the sense of orientation, and we have

$$X_{ij} = -X_{ji}. \tag{5.2.14}$$

The matrices operating on the distinct pairs of coordinates must commute and we can verify straightforwardly that there are the following three vanishing commutators

$$[X_{01}, X_{23}] = 0, \qquad [X_{12}, X_{03}] = 0, \qquad [X_{02}, X_{13}] = 0. \tag{5.2.15}$$

In order to determine the other commutation relations let us notice that in the group $\boldsymbol{SO}(4)$ we can distinguish four subgroups of three-dimensional rotations acting on the spaces G_{ijk} with coordinates x^i, x^j, x^k, which are the following

$$G_{012},\ G_{123},\ G_{230},\ G_{301}. \tag{5.2.16}$$

The infinitesimal matrices of $\boldsymbol{SO}(4)$ can be simultaneously considered as three infinitesimal matrices of each of the subgroups of three-dimensional rotations. We have the following separation

$$\begin{aligned} &G_{012}\colon X_{12},\ X_{20},\ X_{01}; \quad G_{123}\colon X_{23},\ X_{31},\ X_{12}; \\ &G_{230}\colon X_{30}, X_{02}, X_{23}; \quad G_{301}\colon X_{01},\ X_{13},\ X_{30}. \end{aligned} \tag{5.2.17}$$

The three matrices of a subgroup satisfy the commutation relations of the infinitesimal matrices of the group $\boldsymbol{SO}(3)$ given by (3.1.7). Thus we obtain twelve commutation relations which can, on the other hand, be directly verified from the explicit expressions for the matrices (5.2.13). Thus we have, for example,

$$[X_{12}, X_{23}] = X_{31}, \qquad [X_{30}, X_{02}] = X_{23}. \tag{5.2.18}$$

5.2.4 Irreducible Representations

The calculation of the irreducible representations of a continuous group is reduced to obtaining infinitesimal matrices of these representations, as we have seen for the group $\boldsymbol{SO}(3)$. Furthermore, we are going to see that the infinitesimal matrices of the irreducible representations of the group $\boldsymbol{SO}(3)$, given by (3.2.34), allow us to calculate those of $\boldsymbol{SO}(4)$. To do so let us form the following combinations of the infinitesimal matrices

$$\begin{aligned} B_1 &= \tfrac{1}{2}(X_{23} + X_{01}), & C_1 &= \tfrac{1}{2}(X_{23} - X_{01}), \\ B_2 &= \tfrac{1}{2}(X_{31} + X_{02}), & C_2 &= \tfrac{1}{2}(X_{31} - X_{02}), \\ B_3 &= \tfrac{1}{2}(X_{12} + X_{03}), & C_3 &= \tfrac{1}{2}(X_{12} - X_{03}). \end{aligned} \tag{5.2.19}$$

The use of the commutation relations between the infinitesimal matrices X_{ik} allows us to show easily that the matrices B_j all satisfy the same commutation relations as the infinitesimal matrices of the group of three-dimensional rotations, and that the same holds for the matrices C_j. Furthermore, the matrices B_i and C_k commute in pairs.

These commutation relations are satisfied by the following matrices

$$B_i = A_i{}^{(j)} \otimes I_{2j'+1}, \qquad C_i = I_{2j+1} \otimes A_i{}^{(j')}, \qquad i = 1, 2, 3, \tag{5.2.20}$$

where the matrices $A_i{}^{(j)}$ and $A_i{}^{(j')}$ are the infinitesimal matrices of two irreducible representations of the group $\boldsymbol{SO}(3)$. The matrices $I_{2j'+1}$ and I_{2j+1} are, respectively, the unit matrices of dimension $2j'+1$ and $2j+1$. The infinitesimal matrices B_i and C_i are thus both of dimension $(2j'+1)(2j+1)$.

Formulas (5.2.20) show that the group $\boldsymbol{SO}(4)$ is isomorphic to the direct product of two groups of three-dimensional rotations. Since all the irreducible representations of the direct product of two groups can be formed by composition of the irreducible representations of the factor groups, it follows from this that the irreducible representations of $\boldsymbol{SO}(4)$ are identical to the direct products of irreducible representations of two groups of three-dimensional spacial rotations.

We have seen that the irreducible representations of $\boldsymbol{SO}(3)$ are characterized by their weight j which can be a integer of half-integer positive number. Consequently the irreducible representations of $\boldsymbol{SO}(4)$ are characterized by two numbers j and j', each of which can be an integer or half-integer.

Let us introduce the new notations

$$\begin{aligned} A_1{}^{(+)} &= X_{23} & A_2{}^{(+)} &= X_{31} & A_3{}^{(+)} &= X_{12} \\ A_1{}^{(-)} &= X_{01} & A_2{}^{(-)} &= X_{02} & A_3{}^{(-)} &= X_{03}. \end{aligned} \tag{5.2.21}$$

Relations (5.2.19) and (5.2.20) then give us

$$A_i{}^{(+)} = A_i{}^{(j)} \otimes I_{2j'+1} + I_{2j+1} \otimes A_i{}^{(j')},$$

$$A_i^{(-)} = A_i^{(j)} \otimes I_{2j'+1} - I_{2j+1} \otimes A_i^{(j')}, \tag{5.2.22}$$

where $i = 1, 2, 3$ and j, j' are nonnegative integer or half-integer numbers. Relations (5.2.22) give us the matrices of the irreducible representations of $\boldsymbol{SO}(4)$. These representations will be denoted $\Gamma^{(jj')}$. Their dimension is equal to the product of the dimensions of the irreducible representations of $\boldsymbol{SO}(3)$, that is to say, $(2j+1)(2j'+1)$.

5.3 Solved Problems

EXERCISE **5.1**

The transformations of the group $\boldsymbol{O}(4, \mathbb{C})$ may be written in the forms (5.2.2) and (5.2.4), that is to say,

$$x'^i = A^i{}_j x^j, \qquad x'_j = A_j{}^k x_k, \qquad i, j, k = 0, 1, 2, 3.$$

1. Prove the orthogonality relation

$$A^j{}_i A_j{}^k = \delta_i{}^k.$$

2. An infinitesimal transformation of the group $\boldsymbol{O}(4, \mathbb{C})$ in the neighborhood of the identity transformation may be written

$$A_j{}^k = \delta_j{}^k + \varepsilon_j{}^k, \qquad A^j{}_i = \delta^j{}_i + \varepsilon^j{}_i,$$

with $|\varepsilon_j{}^k| \ll 1$. Prove that

$$\varepsilon^k{}_i = -\varepsilon_i{}^k.$$

3. Prove that

$$\varepsilon_{ik} = -\varepsilon_{ki}.$$

4. Deduce from this the number of independent parameters of an infinitesimal transformation of the groups $\boldsymbol{O}(4, \mathbb{C})$ and $\boldsymbol{O}(4, \mathbb{R})$.

SOLUTION **5.1**

1. The transformations of the group $\boldsymbol{O}(4, \mathbb{C})$ preserve the norm of a position vector with components x^j. Thus we have

$$x_i x^i = x'_j x'^j.$$

Substituting the transformation relations for the group into the preceding equation yields

$$x'_j x'^j = (A_j{}^k x_k)(A^j{}_i x^i) = x_i x^i.$$

In order that the latter equality be satisfied we must have

$$A_j{}^k A^j{}_i = \delta_i{}^k.$$

2. Let us use this orthogonality relation for the infinitesimal transformations. This yields

$$\begin{aligned} A_j{}^k = \delta_i{}^k &= (\delta_j{}^k + \varepsilon_j{}^k)(\delta^j{}_i + \varepsilon^j{}_i) \\ &= \delta_j{}^k \delta^j{}_i + \delta_j{}^k \varepsilon^j{}_i + \varepsilon_j{}^k \delta^j{}_i + O(\varepsilon^3) \\ &= \delta^k{}_i + \varepsilon^k{}_i + \varepsilon_i{}^k + O(\varepsilon^2). \end{aligned}$$

   Identifying the terms appearing in this equality gives us, to first order

$$\varepsilon^k{}_i = -\varepsilon_i{}^k.$$

3. Let us write $g_i{}^k$ for the components of the fundamental tensor. We have

$$\varepsilon_j{}^k = g^{km} \varepsilon_{jm}, \qquad \varepsilon^k{}_j = g^{km} \varepsilon_{mj}.$$

   Taking account of the relation $\varepsilon^k{}_i = -\varepsilon_i{}^k$ we obtain

$$\varepsilon_{jm} = -\varepsilon_{mj}.$$

   The infinitesimal quantities ε_{mj} must be antisymmetric.

4. The infinitesimal transformations of the group $\boldsymbol{O}(4, \mathbb{C})$ consist of sixteen parameters ε_{ik}. The quantities ε_{ii} are zero, which leaves twelve nonzero parameters. As these quantities ε_{jm} are antisymmetric there remain six independent parameters. These are complex numbers for the group $\boldsymbol{O}(4, \mathbb{C})$ and real numbers for $\boldsymbol{O}(4, \mathbb{R})$.

Exercise **5.2**

The infinitesimal matrices of the group $\boldsymbol{SO}(4, \mathbb{R})$ are given by (5.2.13).

1. Calculate the commutator $[X_{12}, X_{23}]$.

2. Calculate the commutator $[X_{12}, X_{20}]$.

3. What are the infinitesimal matrices of the subgroup of three-dimensional rotations G_{012} of the group $\boldsymbol{SO}(4, \mathbb{R})$?

4. What are the commutation relations of the infinitesimal matrices of the subgroup G_{012}.

SOLUTION **5.2**

1. The matrices X_{12} and X_{23} are given by (5.2.13). We obtain the following products

$$X_{12}X_{23} = \begin{bmatrix} 0 & 0 & 0 & 0 \\ 0 & 0 & -1 & 0 \\ 0 & 1 & 0 & 0 \\ 0 & 0 & 0 & 0 \end{bmatrix} \begin{bmatrix} 0 & 0 & 0 & 0 \\ 0 & 0 & 0 & 0 \\ 0 & 0 & 0 & -1 \\ 0 & 0 & 1 & 0 \end{bmatrix} = \begin{bmatrix} 0 & 0 & 0 & 0 \\ 0 & 0 & 0 & 1 \\ 0 & 0 & 0 & 0 \\ 0 & 0 & 0 & 0 \end{bmatrix},$$

$$X_{23}X_{12} = \begin{bmatrix} 0 & 0 & 0 & 0 \\ 0 & 0 & 0 & 0 \\ 0 & 0 & 0 & -1 \\ 0 & 0 & 1 & 0 \end{bmatrix} \begin{bmatrix} 0 & 0 & 0 & 0 \\ 0 & 0 & -1 & 0 \\ 0 & 1 & 0 & 0 \\ 0 & 0 & 0 & 0 \end{bmatrix} = \begin{bmatrix} 0 & 0 & 0 & 0 \\ 0 & 0 & 0 & 0 \\ 0 & 0 & 0 & 0 \\ 0 & 1 & 0 & 0 \end{bmatrix}.$$

The commutator $[X_{12}, X_{23}]$ thus has as its expression

$$[X_{12}, X_{23}] = X_{12}X_{23} - X_{23}X_{12} = \begin{bmatrix} 0 & 0 & 0 & 0 \\ 0 & 0 & 0 & 1 \\ 0 & 0 & 0 & 0 \\ 0 & -1 & 0 & 0 \end{bmatrix} = -X_{13} = X_{31}.$$

2. A calculation of the same type give us the commutator

$$[X_{12}, X_{20}] = X_{01}.$$

3. The subgroup G_{012} consists of the rotations carried out in the following planes: the spacio–temporal planes (x^0, x^1), (x^0, x^2); the spacial plane (x^1, x^2). The elements of the subgroup G_{012} are the matrices X_{01}, X_{20}, X_{12}. The commutation relations are thus identical with those of the group $\boldsymbol{SO}(3)$. Thus it is verified, for example, for the commutator $[X_{12}, X_{20}]$, which gives us again an element, X_{01}, of the group.

EXERCISE **5.3**
The matrices given by (5.2.20) have as their expression

$$B_i = A_i{}^{(j)} \otimes I_{2j'+1}, \qquad C_i = I_{2j+1} \otimes A_i{}^{(j')}, \qquad i = 1, 2, 3.$$

Prove that the matrices satisfy the commutation relations of $\boldsymbol{SO}(3)$.

SOLUTION **5.3**
The properties of the tensor product give us

$$\begin{aligned} B_i B_k &= (A_i{}^{(j)} \otimes I_{2j'+1})(A_k{}^{(j)} \otimes I_{2j'+1}) \\ &= (A_i{}^{(j)} A_k{}^{(j)} \otimes I_{2j'+1} I_{2j'+1}), \end{aligned}$$

whence

$$[B_i, B_k] = [A_i{}^{(j)}, A_k{}^{(j)}] \otimes I_{2j'+1}.$$

The matrices $A_i{}^{(j)}$ are the matrices of the irreducible representations of the group $\boldsymbol{SO}(3)$. The commutator $[A_i{}^{(j)}, A_k{}^{(j)}]$ thus satisfies the commutation relations of $\boldsymbol{SO}(3)$, and the same holds for the matrices B_i and C_j.

6

Representations of the Lorentz Groups

6.1 Irreducible Representations

6.1.1 Relations Between the Groups $\boldsymbol{SO}(3,1)^{\uparrow}$ *and* $\boldsymbol{SO}(4)$

We are going to see that in a neighborhood of the unit element the proper Lorentz group $\boldsymbol{SO}(3,1)^{\uparrow}$ is related in a unique way to the group of four-dimensional rotations $\boldsymbol{SO}(4)$.

Let us remark first of all that the number of independent parameters of these two groups is identical. As we have seen previously in the course of Chapter 5, the number of independent parameters of $\boldsymbol{SO}(4)$ is six, and the same is true of the Lorentz group. In fact, the matrices of the generalized Lorentz group satisfying equality (5.1.7), which is equivalent to ten conditions imposed on the matrix elements which are sixteen in number. Thus there remain six independent parameters also.

These six parameters of the Lorentz group can, for example, be the three components of the velocity of motion relative to the frame of reference and the three Euler angles of the rotations defining the orientation with respect to these reference systems. In order to show the relations between the Lorentz group and the groups $\boldsymbol{SO}(4)$ we shall use as parameters the six angles of rotation in the two-dimensional planes defined by (5.2.9); these angles are written in accordance with formulas (5.2.10), that is to say,

$$\varphi_{01},\ \varphi_{02},\ \varphi_{03};\ \varphi_{12},\ \varphi_{13},\ \varphi_{23}, \tag{6.1.1}$$

where the indices denote the corresponding planes.

Rotations in a Plane

Let us consider, for the moment, a transformation of the group of four-dimensional rotations which leave the coordinates x^2 and x^3 unchanged. Thus it is a question of a change of coordinates and, where we write $\varphi = \varphi_{01}$,

$$x'^0 = x^0 \cos\varphi - x^1 \sin\varphi, \qquad x'^1 = x^0 \sin\varphi + x^1 \cos\varphi. \tag{6.1.2}$$

Let us study what relations (6.1.2) become under the transformation of the variable x^0 and φ defined by

$$x^0 \longrightarrow ix^0, \qquad x'^0 \longrightarrow ix'^0, \qquad \varphi \longrightarrow i\varphi.$$

We obtain

$$ix'0 = ix^0 \cos i\varphi - x^1 \sin i\varphi, \qquad x'^1 = ix^0 \sin i\varphi + x^1 \cos i\varphi, \tag{6.1.3}$$

that is

$$x'^0 = x^0 \cosh\varphi + x^1 \sinh\varphi, \qquad x'^1 = x^0 \sinh\varphi + x^1 \cosh\varphi. \tag{6.1.4}$$

The rotation (6.1.2), carried out in a plane of $\mathbb{R}^4$, admits the invariant

$$(x^0)^2 + (x^1)^2 = \text{const.}, \tag{6.1.5}$$

whilst the transformation (6.1.4) admits as invariant

$$-(x^0)^2 + (x^1)^2 = \text{const.} \tag{6.1.6}$$

The latter is deduced from (6.1.5) by changing x^0 into ix^0. This invariant quadratic form is, as a matter of fact, the invariant of the Lorentz transformation in the plane (x^0, x^1). The transformation (6.1.4) it thus a Lorentz two-dimensional transformation. Analogously we can obtain the relations between the Lorentz transformations and those of the group $\boldsymbol{SO}(4)$ in the spacio–temporal planes (x^0, x^2) and (x^0, x^3), and which are obtained from formulas of the type of (6.1.4).

As for the rotations in the planes (x^1, x^2), (x^1, x^3), and (x^2, x^3), these coincide with the rotations of the group of spacial rotations $\boldsymbol{SO}(3)$ which is a subgroup of $\boldsymbol{SO}(4)$.

Relations Between $\boldsymbol{SO}(3,1)^\uparrow$ and $\boldsymbol{SO}(4)$

Finally, the Lorentz transformations can be obtained from the four-dimensional rotations by changing the real parameters of the rotations φ_{01} $(i = 1, 2, 3)$ in the spacio–temporal planes (x^0, x^1) into the pure imaginary quantities $i\varphi_{0i}$, and at the same time changing x^0 into ix^0. Thus we obtain a correspondence between each transformation of the two groups.

However, the matrix elements of the rotation matrices of $\boldsymbol{SO}(4)$ are sines or cosines, whilst those of the matrices of the Lorentz transformations are not periodic functions. Consequently a unique relation between the transformations of $\boldsymbol{SO}(4)$ and those of $\boldsymbol{SO}(3)^\dagger$ only hold in a neighborhood of the unit element.

Let us write $R(\varphi_{01}, \varphi_{02}, \varphi_{03}; \varphi_{12}, \varphi_{13}, \varphi_{23})$ for the matrix of a rotation of $\boldsymbol{SO}(4)$ and K for the following matrix

$$K = \begin{bmatrix} i & 0 & 0 & 0 \\ 0 & 1 & 0 & 0 \\ 0 & 0 & 1 & 0 \\ 0 & 0 & 0 & 1 \end{bmatrix}. \tag{6.1.7}$$

The matrix $\Lambda(\varphi_{01}, \varphi_{02}, \varphi_{03}; \phi_{12}, \phi_{13}, \phi_{23})$ of the Lorentz group which corresponds to the rotation R can then be written in the form of the product

$$\Lambda(\varphi_{01}, \varphi_{02}, \varphi_{03}; \phi_{12}, \phi_{13}, \phi_{23}) = K^{-1} R(i\varphi_{01}, i\varphi_{02}, i\varphi_{03}; \phi_{12}, \phi_{13}, \phi_{23}) K. \tag{6.1.8}$$

6.1.2 *Infinitesimal Matrices*

According to formulas (6.1.4) the rotation matrix in the plane (x^0, x^1) is

$$M_{01} = \begin{bmatrix} \cosh\varphi_{01} & \sinh\varphi_{01} & 0 & 0 \\ \sinh\varphi_{01} & \cosh\varphi_{01} & 0 & 0 \\ 0 & 0 & 1 & 0 \\ 0 & 0 & 0 & 1 \end{bmatrix}.$$

The rotation matrices in the spacio–temporal planes (x^0, x^1) and (x^0, x^3) are matrices analogous to those given by formulas (5.2.12), where we replace the circular functions by the corresponding positive hyperbolic functions. The infinitesimal matrices for the spacio–temporal rotations are then the following:

$$Y_{01} = \begin{bmatrix} 0 & 1 & 0 & 0 \\ 1 & 0 & 0 & 0 \\ 0 & 0 & 0 & 0 \\ 0 & 0 & 0 & 0 \end{bmatrix}, \qquad Y_{02} = \begin{bmatrix} 0 & 0 & 1 & 0 \\ 0 & 0 & 0 & 0 \\ 1 & 0 & 0 & 0 \\ 0 & 0 & 0 & 0 \end{bmatrix}, \tag{6.1.9a}$$

$$Y_{03} = \begin{bmatrix} 0 & 0 & 0 & 1 \\ 0 & 0 & 0 & 0 \\ 0 & 0 & 0 & 0 \\ 1 & 0 & 0 & 0 \end{bmatrix}. \tag{6.1.9b}$$

The infinitesimal matrices corresponding to the spacial rotations, denoted Y_{12}, Y_{13}, Y_{23}, are identical to the matrices X_{12}, X_{13}, X_{23} of the group $\boldsymbol{SO}(4)$ given by (5.2.13).

The Matrix of an Infinitesimal Transformation

The six infinitesimal matrices of the Lorentz group are linearly independent and are able to form a basis of the six-dimensional vector space E_6. We prove that all the infinitesimal transformations of the Lorentz group have matrices which can be expressed in the basis of the group's infinitesimal matrices in the form

$$Y = \varepsilon_{10} Y_{01} + \varepsilon_{02} Y_{02} + \varepsilon_{03} Y_{03} + \varepsilon_{12} Y_{12} + \varepsilon_{13} Y_{13} + \varepsilon_{23} Y_{23}. \tag{6.1.10}$$

Taking account of the expressions for the infinitesimal matrices, an infinitesimal Lorentz transformation may then be written in the matrix form

$$\begin{bmatrix} x'^0 \\ x'^1 \\ x'^2 \\ x'^3 \end{bmatrix} = \begin{bmatrix} x^0 \\ x^1 \\ x^2 \\ x^3 \end{bmatrix} + \begin{bmatrix} 0 & \varepsilon_{01} & \varepsilon_{02} & \varepsilon_{03} \\ \varepsilon_{01} & 0 & -\varepsilon_{12} & -\varepsilon_{13} \\ \varepsilon_{02} & \varepsilon_{12} & 0 & -\varepsilon_{23} \\ \varepsilon_{03} & \varepsilon_{13} & \varepsilon_{23} & 0 \end{bmatrix} \begin{bmatrix} x^0 \\ x^1 \\ x^2 \\ x^3 \end{bmatrix}. \tag{6.1.11}$$

Commutation Relations

In Chapter 4 we have studied the commutation relations of the infinitesimal matrices of $\boldsymbol{SO}(4)$. According to (6.1.8) we obtain the following relation between the infinitesimal matrices of the Lorentz group and those of the group of four-dimensional rotations

$$\begin{aligned} \left(\frac{\partial \Lambda}{\partial \varphi}\right)_0 &= iK^{-1} \left(\frac{\partial R}{\partial \varphi}\right)_0 K \qquad \text{where} \quad \varphi = \varphi_{01}, \varphi_{02}, \varphi_{03}, \\ \left(\frac{\partial \Lambda}{\partial \phi}\right)_0 &= K^{-1} \left(\frac{\partial R}{\partial \phi}\right)_0 K \qquad \text{where} \quad \phi = \varphi_{12}, \varphi_{13}, \varphi_{23}. \end{aligned} \tag{6.1.12}$$

Since a similarity transformation of the type $\Lambda \to K^{-1}\Lambda K$ does not modify the commutation relations, it follows that these relations between the infinitesimal matrices of the Lorentz group will be the same as for the matrices $iX_{01}, iX_{02}, iX_{03}, X_{12}, X_{13}, X_{23}$, where the matrices X_{ik} are the infinitesimal matrices of the group $\boldsymbol{SO}(4)$ and of which the properties have been studied in the course of the preceding Chapter 5.

These commutation properties are also satisfied by the infinitesimal matrices of an arbitrary representation of the Lorentz group. We have already studied these properties for the group $\boldsymbol{SO}(3)$, and they are related to Sophus Lie's theorem given in relation (2.1.26).

6.1.3 Irreducible Representations

The calculation of the irreducible representations of a continuous group reduces to obtaining the infinitesimal matrices of these representations. Now, the infinitesimal matrices of the irreducible representations of the Lorentz group are going to coincide with the corresponding matrices of the group $\boldsymbol{SO}(4)$ as far as the spacial rotations are concerned, or they differ by only a factor of i for the spacio–temporal rotations. The matrices of the irreducible representations of $\boldsymbol{SO}(4)$ are given by formulas (5.2.22).

For the matrices of the Lorentz group let us use notations analogous to those used for the group $\boldsymbol{SO}(4)$. Let us write B_{ij} for the infinitesimal matrices of the irreducible representations of the Lorentz group, where the indices i, j are related to each of the planes (x^i, x^j). Let us introduce new notations similar to those of (5.2.21)

$$\begin{aligned} &{B_1}^{(+)} = B_{23}, \qquad {B_2}^{(+)} = B_{31}, \qquad {B_3}^{(+)} = B_{12}, \\ &{B_1}^{(-)} = B_{01}, \qquad {B_2}^{(-)} = B_{02}, \qquad {B_3}^{(-)} = B_{03}. \end{aligned} \tag{6.1.13}$$

Expressions (5.2.22) for the matrices of the irreducible representations of $\boldsymbol{SO}(4)$ then give us the corresponding matrices of the Lorentz group

$$\begin{aligned} {B_i}^{(+)} &= {A_i}^{(j)} \otimes I_{2j'+1} + I_{2j+1} \otimes {A_i}^{(j')}, \\ {B_i}^{(-)} &= i{A_i}^{(j)} \otimes I_{2j'+1} - iI_{2j+1} \otimes {A_i}^{(j')}, \end{aligned} \tag{6.1.14}$$

where $i = 1, 2, 3$ and j, j' are nonnegative integers or half-integers. Let us recall that the matrices ${A_i}^{(j)}$ and ${A_i}^{(j')}$ are the infinitesimal matrices of two irreducible representations of the group $\boldsymbol{SO}(3)$ of respective weights j and j'. The matrices of the irreducible representations of the Lorentz group are thus specified by a pair of numbers, j and j', and these representations are denoted $\Gamma^{(jj')}$. The dimension of the representation is equal to $(2j + 1)(2j' + 1)$.

As a result of the presence of the imaginary number i which appears in the second formula of (6.1.14), the matrices ${B_i}^{(-)}$ are not anti-Hermitian, and consequently the finite-dimensional representations $\Gamma^{(jj')}$ are not unitary. This is related to the Lorentz group not being compact.

In addition to these finite-dimensional irreducible representations the Lorentz group also admits infinite-dimensional unitary irreducible representations. However, since these representations have only one quite limited domain of application we shall not study them.

6.2 The Group $\boldsymbol{SL}(2, \mathbb{C})$

6.2.1 *Two-Component Spinors*

The group $\boldsymbol{SL}(2, \mathbb{C})$ is the group of linear transformations which, when operating on a vector (ψ^1, ψ^2) with two complex components, transforms it into a vector (ψ'^1, ψ'^2) according to the relations

$$\psi'^1 = a\psi^1 + b\psi^2, \qquad \psi'^2 = c\psi^1 + d\psi^2, \tag{6.2.1}$$

where the parameters a, b, c, d are complex numbers related by the conditions

$$\begin{vmatrix} a & b \\ c & d \end{vmatrix} = ad - cb = 1. \tag{6.2.2}$$

The group $\boldsymbol{SL}(2, \mathbb{C})$ is thus a unimodular group. Condition (6.2.2) shows that the complex parameters can be determined as functions of three others. Thus we have only six independent real parameters which define a transformation of the group $\boldsymbol{SL}(2, \mathbb{C})$. The matrix of the transformations can be written in the form

$$M(SL) = \begin{vmatrix} a & b \\ c & \dfrac{1+cb}{a} \end{vmatrix}. \tag{6.2.3}$$

As we have seen for the group $\boldsymbol{SU}(2)$, which allows us to define the spinors of three-dimensional space, the group $\boldsymbol{SL}(2, \mathbb{C})$ is going to allow us to define the spinors of four-dimensional space.

By definition the vectors $\chi = (\psi^1, \psi^2)$ with complex components which transform under the transformations of the group $\boldsymbol{SL}(2, \mathbb{C})$ are called *spinors of four-dimensional space.*

If this space is considered as being the space–time of Relativity, it will be said that they are spinors of space–time or spinors of Minkowski space.

'Dotted' Spinors

Let us write the complex conjugates of the numbers ψ^1 and ψ^2 in the following form

$$\psi^{1*} = \psi^{\dot{1}}, \qquad \psi^{2*} = \psi^{\dot{2}}. \tag{6.2.4}$$

The indices with a dot over them form a classical notation for representing the complex conjugates of spinors, and are called DOTTED SPINORS.

Considering the complex conjugate quantities, the transformations (6.2.1) are then written

$$\psi'^{\dot{1}} = a^*\psi^{\dot{1}} + b^*\psi^{\dot{2}}, \qquad \psi'^{\dot{2}} = c^*\psi^{\dot{1}} + d^*\psi^{\dot{2}}. \tag{6.2.5}$$

As a result of the independence of the different complex parameters a and a^*, b and b^*, c and c^*, the dotted spinors $\dot{\chi} = (\psi^{\dot{1}}, \psi^{\dot{2}})$ are going to transform differently from the spinors $\chi = (\psi^1, \psi^2)$. The matrices of the transformations of dotted spinors are given by

$$\dot{M}(SL) = \begin{bmatrix} a^* & b^* \\ c^* & \dfrac{1 + c^*b^*}{a^*} \end{bmatrix}. \tag{6.2.6}$$

Thus in relativistic theory there exist two types of spinor, those which transform under (6.2.3) and those under (6.2.6).

(The spinors $\phi = (\psi^1, \psi^2)$ and $\eta = (\psi^{\dot{1}}, \psi^{\dot{2}})$ correspond, respectively, to the representations $(\frac{1}{2}, 0)$ and $(0, \frac{1}{2})$ of the Lorentz group (see Section 6.3.2)). These spinors are actually written $\phi = \phi_{\mathrm{R}}$ and $\eta = \eta_R ML$, where R stands for 'right handed' and L for 'left handed'.)

Contravariant and Covariant Components

We can consider the components ψ^1 and ψ^2 as the 'contravariant' components of a spinor. In order to introduce some algebraic properties analogous to those of vectors we can use the 'metric spinor' g_{ij}, which has as its matrix

$$g_{ij} = \begin{bmatrix} 0 & 1 \\ -1 & 0 \end{bmatrix}. \tag{6.2.7}$$

The 'covariant' components ψ_i of a spinor are then related to the covariant components ψ^k by

$$\psi_i = g_{ik}\psi^k, \tag{6.2.8}$$

so that we have

$$\psi_1 = \psi^2, \qquad \psi_2 = -\psi^1. \tag{6.2.9}$$

For the components of the dotted spinors we have the same relations

$$\psi_{\dot{1}} = \psi^{\dot{2}}, \qquad \psi_{\dot{2}} = -\psi^{\dot{1}}. \tag{6.2.10}$$

6.2.2 *Higher-Order Spinors*

The spinors with two components are spinors of order 1. Spinors of higher order are 1 defined as having components which transform as the products of the components of several spinors of order 1.

The indices of a higher-order spinor can be dotted for some and not for others. For example, we obtain three distinct types of second-order spinors formed from spinors of order 1, that is to say,

$$\zeta^{\alpha\beta} \sim \psi^\alpha \phi^\beta, \qquad \chi^{\alpha\dot{\beta}} \sim \psi^\alpha \eta^{\dot{\beta}}, \qquad \lambda^{\dot{\alpha}\dot{\beta}} \sim \eta^{\dot{\alpha}} \mu^{\dot{\beta}}, \tag{6.2.11}$$

where the symbol '$\sim$' means 'transforms like'. Indicating only the total order of a spinor therefore does not suffice for specifying the type of the spinor. We must also indicate the order of the spinor by a pair of numbers (k, k'), which respectively represent the number of undotted indices and dotted indices.

Quite generally, we can therefore construct monomials analogous to those we have studied for the group $\boldsymbol{SU}(2)$, formed from the spinors of order 1 (ψ_1, ψ_2) and $(\eta_{\dot{1}}, \eta_{\dot{2}})$. We obtain the following general expression

$$P_{kk'} = (\psi_1)^{\nu-k} (\eta_{\dot{1}})^{\nu'-k'} (\psi_2)^k (\eta_{\dot{2}})^{k'}. \tag{6.2.12}$$

It is a monomial of degree $\nu + \nu'$ with $0 \leqslant k \leqslant \nu$ and $0 \leqslant k' \leqslant \nu'$. These monomials can be considered as the components of a spinor. For given ν and ν' we have $(\nu + 1)(\nu' + 1)$ monomials, that is to say,

$$\begin{gathered} P_{00} = {\psi_1}^\nu {\eta_{\dot{1}}}^{\nu'}, \qquad P_{01} = {\psi_1}^\nu {\eta_{\dot{1}}}^{\nu'-1} \psi_2, \qquad P_{10} = {\psi_1}^{\nu-1} {\eta_{\dot{1}}}^{\nu'} \psi_2, \\ \vdots \\ P_{\nu,\nu'-1} = \eta_{\dot{1}} {\psi_2}^\nu {\eta_{\dot{2}}}^{\nu'-1}, \qquad P_{\nu\nu'} = {\psi_1}^\nu {\eta_{\dot{2}}}^{\nu'}. \end{gathered} \tag{6.2.13}$$

Let us introduce the classical notations $j = \nu/2$ and $j' = \nu'/2$. The vectors with these $(2j+1)(2j'+1)$ monomials as components are called SPINORS OF RANK $(2j+1)(2j'+1)$ of four-dimensional space.

Respectively subjecting the components ψ_1, ψ_2 and $\eta_{\dot{1}}, \eta_{\dot{2}}$ to the transformations (6.2.1) and (6.2.5), for the transformation of the monomial $P_{kk'}$ given by (6.2.12) we obtain

$$\begin{aligned} P_{kk'} &= (a\psi_1 + b\psi_2)^{\nu-k} (a^* \eta_{\dot{1}} + b^* \eta_{\dot{2}})^{\nu'-k'} \\ &\quad \times (c\psi_1 + d\psi_2)^k (c^* \eta_{\dot{1}} + d^* \eta^{\dot{2}})^{k'}. \end{aligned} \tag{6.2.14}$$

This is the general transformation formula for the components of spinors of rank $(2j+1)(2j'+1)$.

6.2.3 Representations of the Groups $\boldsymbol{SL}(2, \mathbb{C})$

Two-Dimensional Representations

The two-component spinors (ψ_1, ψ_2) form a vector space which is able to serve as a representation space of the group $\boldsymbol{SL}(2, \mathbb{C})$. The matrices (6.2.3) form a matrix representation of this group.

On considering the notations above, the spinors (ψ_1, ψ_2) are characterized by the values $j = \frac{1}{2}$ and $j' = 0$. Such a representation will be denoted $D^{(1/2,0)}$.

Similarly the dotted spinors $(\eta_{\dot{1}}, \eta_{\dot{2}})$ form a representation denoted $D^{(0,1/2)}$, the matrices of which are given by (6.2.6).

Four-Dimensional Representations

Let us calculate the matrices of the representations $D^{(1/2,1/2)}$ corresponding to $j = \frac{1}{2}$ and $j' = \frac{1}{2}$. It is a representation of dimension four, the components of the spinors of order $(\frac{1}{2}, \frac{1}{2})$ are

$$P_{00} = \psi_1\eta_{\dot{1}}, \qquad P_{01} = \psi 1\eta_{\dot{2}}, \qquad P_{10} = \psi_2\eta_{\dot{1}}, \qquad P_{11} = \psi_2\eta_{\dot{2}}. \tag{6.2.15}$$

Let us apply to (ψ_1, ψ_2) and $(\eta_{\dot{1}}, \eta_{\dot{2}})$ the transformations (6.2.1) and (6.2.5), respectively, and let us substitute these transformed quantities in the monomials (6.2.15). We obtain

$$\begin{aligned} P'_{00} &= (a\psi_1 + \psi_2)(a^*\eta_{\dot{1}} + b^*\eta_{\dot{2}}) \\ &= aa^*P_{00} + ab^*P_{01} + ba^*P_{10} + bb^*P_{11}, \\ P'_{01} &= ac^*P_{00} + ad^*P_{01} + bc^*P_{10} + bd^*P_{11}, \\ P'_{10} &= ca^*P_{00} + cb^*P_{01} + da^*P_{10} + db^*P_{11}, \\ P'_{11} &= cc^*P_{00} + cd^*P_{01} + dc^*P_{10} + dd^*P_{11}. \end{aligned} \tag{6.2.16}$$

Let us remark that the transformations suffered by $P_{00}, P_{01}, P_{10}, P_{11}$ again give the elements of the initial vector space formed by the $(\frac{1}{2}, \frac{1}{2})$ spinors. This clearly results from the way in which they have been formed. The matrix of the transformation is

$$M(D^{(1/2,1/2)}) = \begin{bmatrix} aa^* & ab^* & ba^* & bb^* \\ ac^* & ad^* & bc^* & bd^* \\ ca^* & cb^* & da^* & db^* \\ cc^* & cd^* & dc^* & dd^* \end{bmatrix}. \tag{6.2.17}$$

These matrices constitute the matrix representation $D^{(1/2,1/2)}$ of the group $\boldsymbol{SL}(2,\mathbb{C})$. The matrix (6.2.17) can be written in the form of the tensor product of two matrices

$$M(D^{(1/2,1/2)}) = \begin{bmatrix} a & b \\ c & d \end{bmatrix} \otimes \begin{bmatrix} a^* & b^* \\ c^* & d^* \end{bmatrix}. \tag{6.2.18}$$

The first matrix is that of the representation $D^{(1/2,0)}$, and the second that of the representation $D^{(0,1/2)}$. The representation $D^{(1/2,1/2)}$ can thus be decomposed into the form of a direct product of two representations, that is,

$$D^{(1/2,1/2)} = D^{(1/2,0)} \otimes D^{(0,1/2)}. \tag{6.2.19}$$

Arbitrary-Dimensional Representations

Let us consider, for the moment, the general expression for spinors of rank $(2j+1)(2j'+1)$, that is,

$$P_{kk'} = (\psi_1)^{2j-k}(\eta_{\dot{1}})^{2j'-k'}(\psi_2)^k(\eta_{\dot{2}})^{k'}, \tag{6.2.20}$$

with $k = 0, \dots, 2j$ and $k' = 0, \dots, 2j'$. Let us apply the transformations (6.2.1) and (6.2.5) respectively to (ψ_1, ψ_2) and $(\eta_{\dot{1}}, \eta_{\dot{2}})$. We obtain

$$\begin{aligned} P'_{kk} &= (a\psi_1 + b\psi_2)^{2j-k}(a^*\eta_{\dot{1}} + b^*\eta_{\dot{2}})^{2j'-k'}(c\psi_1 + d\psi_2)^k(c^*\eta_{\dot{1}} + d^*\eta_2)^{k'} \\ &= \sum_{p'=0}^{2j'}\sum_{p=0}^{2j} S^{(jj')}_{kk',pp'}P_{pp'}. \end{aligned} \tag{6.2.21}$$

The $(2j+1)(2j'+1)$ monomials (6.2.20) may all undergo between themselves a linear transformation, thanks to the way in which they have been formed. The matrices of this transformation constitute a matrix representation of dimension $(2j+1)(2j'+1)$ of the group $\boldsymbol{SL}(2,\mathbb{C})$.

6.2.4 Irreducible Representations

Operations on Spinors

The operation called symmetrization of a spinor consists in calculating an average over all possible permutations of the set of its indices. If we write, for example, (ijk) for this operation on the indices i, j, k of a spinor, then by definition we have

$$\psi_{(ijk)} = \frac{1}{3!}(\psi_{ijk} + \psi_{ikj} + \psi_{jik} + \psi_{jki} + \psi_{kij} + \psi_{kji}). \tag{6.2.22}$$

A spinor is said to be symmetric over a set of indices if it does not change under the operation of symmetrization applied to the set of its indices, that is,

$$\psi_{\dots ijk\dots} = \psi_{\dots(ijk)\dots}. \tag{6.2.23}$$

The operations on spinors are analogous to those on tensors. However, the contraction of spinors can be carried out only on pairs of indices of the same kind; either on two dotted indices or on two undotted indices, since the summation over indices of different kinds is not an invariant operation. Furthermore, contraction over a pair of indices in which a spinor is symmetric gives a zero value.

The Basis of Irreducible Representations

From the preceding considerations we deduce that the spinor with components

$$\psi_{\alpha_1 \dots \alpha_k, \dot{\beta}_1 \dots \dot{\beta}_m} \tag{6.2.24}$$

symmetric over all the k undotted indices and over all the m dotted indices, cannot serve to form a spinor of lower order. In other words, no linear combination of the different components (6.2.24) can be allowed to become a smaller number of components transforming amongst themselves like those of a spinor in the transformations of the group $\boldsymbol{SL}(2, \mathbb{C})$. The symmetric spinors thus constitute a representation basis of the irreducible representations of $\boldsymbol{SL}(2, \mathbb{C})$.

In particular, let us consider the symmetric spinors of rank $(2j+1) \times (2j'+1)$ formed from the spinors (ψ_1, ψ_2) and $(\psi_{\dot{1}}, \psi_{\dot{2}})$. Let us form spinors of which the components have the form (6.2.20), that is,

$$P_{kk'} = (\psi_1)^{2j-k} (\psi_{\dot{1}})^{2j'-k'} (\psi_2)^k (\psi_{\dot{2}})^{k'}, \tag{6.2.25}$$

with $k = 0, \dots, 2j$ and $k' = 0, \dots, 2j'$. These spinors realize irreducible representations of dimension $(2j+1)(2j'+1)$ of the group $\boldsymbol{SL}(2, \mathbb{C})$. Each representation, denoted $D^{(jj')}$, is specified by a pair of number (j, j'). We prove that this process gives *all* the irreducible representations of the group $\boldsymbol{SL}(2, \mathbb{C})$.

6.3 Spinor Representations of the Lorentz Group

We are going to see examples showing that the irreducible representations $D^{(jj')}$ of the group $\boldsymbol{SL}(2, \mathbb{C})$ constitute representations equivalent to the irreducible representations $\Gamma^{(jj')}$ of the Lorentz group when the sums $j+j'$ are integers. When $j+j'$ are half-integers we have other representations called pure spinor representations.

As the spinors of rank $(2j+1)(2h'+1)$ then form bases for the representations of the Lorentz group we shall say that they are *spinor representations* of this group.

6.3.1 Four-Dimensional Irreducible Representations

Let us consider the representation $D^{(1/2,1/2)}$ of the group $\boldsymbol{SL}(2, \mathbb{C})$ which is based on the spinor with components

$$P_{00} = \psi_1 \psi_{\dot{1}} \qquad P_{01} = \psi_1 \psi_{\dot{2}} \qquad P_{10} = \psi_2 \psi_{\dot{1}} \qquad P_{11} = \psi_2 \psi_{\dot{2}} \tag{6.3.1}$$

The matrices of this representation are given by (6.2.17). As a result of the relations $ad - bc = 1$ and $a^*d^* - b^*c^* = 1$ we have the following equality

$$-P'_{00} P'_{11} + P'_{01} P'_{10} = -P_{00} P_{11} + P_{01} P_{10}. \tag{6.3.2}$$

Let us show that under a change of basis the spinors (6.3.1) form a representation basis of the Lorentz group. Let us note that this method is analogous to that already used for the spinor representation of the spacial rotations $\boldsymbol{SO}(3)$. To do so let us consider the following linear combinations

$$\begin{bmatrix} Q_1 \\ Q_2 \\ Q_3 \\ Q_4 \end{bmatrix} = \frac{1}{\sqrt{2}} \begin{bmatrix} 0 & 1 & 1 & 0 \\ 0 & i & -i & 0 \\ 1 & 0 & 0 & -1 \\ i & 0 & 0 & i \end{bmatrix} \begin{bmatrix} P_{00} \\ P_{01} \\ P_{10} \\ P_{11} \end{bmatrix}$$

$$= \frac{1}{\sqrt{2}} AP, \tag{6.3.3}$$

where the matrix $(1/\sqrt{2})A$ is the matrix of the change of coordinates. The inverse transformation give us in matrix form

$$P = \sqrt{2} A^{-1} Q, \tag{6.3.4}$$

that is, explicitly,

$$P_{00} = Q_3 - iQ_4, \quad P_{01} = Q_1 - iQ_2, \quad P_{10} = Q_1 + iQ_2, \quad P_{11} = -Q_3 - iQ_4.$$

Substituing these expression into the right-hand side of (6.3.2) we obtain

$$-P_{00}P_{11} + P_{01}P_{10} = {Q_1}^2 + {Q_2}^2 + {Q_3}^2 + {Q_4}^2. \tag{6.3.5}$$

On the other hand, the matrix relation (6.3.4) also gives $Q' = (1/\sqrt{2})AP'$ for the transforms and thus a relation similar to (6.3.5) for the transformed quantities. Taking into account (6.3.5) and its analogue for the transforms, relation (6.3.2) may then be written

$${Q'_1}^2 + {Q'_2}^2 + {Q'_3}^2 + {Q'_4}^2 = {Q_1}^2 + {Q_3}^2 + {Q_3}^2 + {Q_4}^2. \tag{6.3.6}$$

The latter relation corresponds to the invariance of a vector which has Q_1, Q_2, Q_3, Q_4 as its components. Since the dotted spinors correspond to the conjugate complex quantities, it is easily verified that Q_1, Q_2, Q_3 are real numbers whilst Q_4 is complex. Thus we obtain quantities Q_i of which the rules of transformation are those of the components of a four-vector.

Furthermore, taking account of the transformation matrix (6.2.17) of a spinor the transformed vector may be written

$$Q' = \frac{1}{\sqrt{2}} AP' = \frac{1}{\sqrt{2}} AM(D^{(1/2,1/2)})P$$

$$= AM(D^{(1/2,1/2)})A^{-1}Q. \tag{6.3.7}$$

Taking (6.2.18) into account, the determinant of the preceding transformation has the value

$$\begin{aligned}\det[AM(D^{(1/2,1/2)})A^{-1}] &= \det M(D^{(1/2,1/2)})\\ &= \det M(D^{(1/2)}) \otimes \dot{M}(D^{(1/2)})\\ &= \det M(D^{(1/2)}) \det \dot{M}(D^{(1/2)})\\ &= 1. \end{aligned} \tag{6.3.8}$$

On the other hand, the matrix element $[AM(D^{(1/2,1/2)})A^{-1}]_{44}$ is written

$$\begin{aligned}[AM(D^{(1/2,1/2)})A^{-1}]_{44} &= \tfrac{1}{2}(|a|^2 + |b|^2 + |c|^2 + |d|^2)\\ &\geqslant 1. \end{aligned} \tag{6.3.9}$$

We thus verify that all the properties of a proper Lorentz transformation are satisfied. Finally, the transformation $AM(D^{(1/2,1/2)})A^{-1}$ transforms the vector (Q_1, Q_2, Q_3, Q_4) analogously to a proper transformation of the Lorentz group acting on a four-vector.

The matrices $AM(D^{(1/2,1/2)}A^{-1}$ thus form a representation of the Lorentz group equivalent to the representation $D^{(1/2,1/2)}$ of the group $\boldsymbol{SL}(2, \mathbb{C})$.

The example set out is a particular case of the representations $D^{(jj')}$ of the group $\boldsymbol{SL}(2, \mathbb{C})$ when $j + j'$ is an *integral* number. In this case the representations $D^{(jj')}$ are the representations equivalent to tensor representations $\Gamma^{(jj')}$ of the proper Lorentz group.

6.3.2 *Two-Dimensional Representations*

Let us study the relations between the infinitesimal matrices of the Lorentz group and those of the group $\boldsymbol{SL}(2, \mathbb{C})$ in the case of two-dimensional representations. To do so let us consider, for example, a rotation through an angle φ in the plane (x_1, x_2). We have the following transforms of the coordinates (x_0, x_1, x_2, x_3) in an orthonormal basis

$$\begin{aligned} &{x'}_0 = x_0, \qquad {x'}_1 = x_1 \cos\varphi + x_2 \sin\varphi,\\ &{x'}_2 = -x_1 \sin\varphi + x_2 \cos\theta, \qquad {x'}_3 = x_3. \end{aligned} \tag{6.3.10}$$

Let us carry out the following linear combinations

$$\begin{aligned} &{x'}_3 - i{x'}_0 = x_3 - ix_0, \qquad {x'}_1 - i{x'}_2 = e^{i\varphi}(x_1 - ix_2),\\ &{x'}_1 + i{x'}_2 = e^{-i\varphi}(x_1 + ix_2), \qquad -{x'}_3 - i{x'}_0 = -x_3 - ix_0. \end{aligned} \tag{6.3.11}$$

These transformations acting on a four-vector correspond to transformations on the components of the spinor (6.3.1) which are obtained by comparing (6.3.11) and (6.3.4), whence,

$$P'_{00} = P_{00}, \quad P'_{01} = e^{i\varphi}P_{01}, \quad P'_{10} = e^{-i\varphi}P_{10}, \quad P'_{11} = P_{11}. \tag{6.3.12}$$

The latter expressions are a particular case of the general transformation (6.2.16). By identifying the coefficients of these two sets of equations we obtain

$$aa^* = 1, \qquad ab^* = bb^* = 0, \qquad ac^* = 0, \qquad ad^* = e^{i\varphi}, \quad \text{etc.} \tag{6.3.13}$$

The solution of this system of equations is

$$a = \pm e^{i\varphi/2}, \qquad b = 0, \qquad c = 0, \qquad d = \pm e^{-i\varphi/2}.$$

The rotation in the plane (x^1, x^2) through an angle φ corresponds to a transformation, also called a rotation by analogy, in the space of $(\frac{1}{2}, \frac{1}{2})$ spinors defined by the matrix:

$$M(D^{(1/2,1/2)}) = \pm \begin{bmatrix} e^{i\varphi/2} & 0 \\ 0 & e^{-i\varphi/2} \end{bmatrix} \otimes \begin{bmatrix} e^{-i\varphi/2} & 0 \\ 0 & e^{i\varphi/2} \end{bmatrix}. \tag{6.3.14}$$

The Representation $\Gamma^{1/2,0}$

According to (6.2.18) the first matrix corresponds to the $D^{(1/2,0)}$ representation and the second to the $D^{(0,1/2)}$ representation of the dotted spinors. Let us calculate the infinitesimal matrix of the representation $D^{(1/2,0)}$. We obtain

$$Z(D^{(1/2,0)}) = \pm\frac{i}{\sqrt{2}} \begin{bmatrix} 1 & 0 \\ 0 & -1 \end{bmatrix}. \tag{6.3.15}$$

Let us use expressions (6.1.14) for the infinitesimal matrices of the representations $\Gamma^{(jj')}$ of the Lorentz group to compare them with the matrices of the representations $D^{(jj')}$ of the group $\boldsymbol{SL}(2, \mathbb{C})$. The infinitesimal matrices of the representations $\Gamma^{(jj')}$ are expressed with the help of the Pauli matrices, the expressions for which we recall

$$\sigma_1 = \begin{bmatrix} 0 & 1 \\ 1 & 0 \end{bmatrix}, \qquad \sigma_2 = i\begin{bmatrix} 0 & -1 \\ 1 & 0 \end{bmatrix}, \qquad \sigma_3 = \begin{bmatrix} 1 & 0 \\ 0 & -1 \end{bmatrix}. \tag{6.3.16}$$

The infinitesimal matrices of the representation $\Gamma^{(1/2,0)}$ of the Lorentz group are then written

$$\begin{aligned} B_1{}^{(+)} &= \tfrac{1}{2}i\sigma_1, & B_2{}^{(+)} &= -\tfrac{1}{2}i\sigma_2, & B_3{}^{(+)} &= \tfrac{1}{2}i\sigma_3, \\ B_1{}^{(-)} &= -\tfrac{1}{2}\sigma_1, & B_2{}^{(-)} &= \tfrac{1}{2}\sigma_2, & B_3{}^{(-)} &= -\tfrac{1}{2}\sigma_3. \end{aligned} \tag{6.3.17}$$

According to notation (6.1.13) we have ${B_3}^{(+)} = B_{12}$. The latter matrix is the one which corresponds to a rotation in the plane (x_1, x_2). We then notice that the matrix $Z(D^{(1/2,0)})$ given by (6.3.15) is identical to ${B_3}^{(+)}$ given by (6.3.17). However, we remark that to a rotation in the usual sense there correspond two transformations in the space of spinors, as has already been seen for the group $\boldsymbol{SO}(3)$.

A similar calculation can be carried out for each of the rotations in a plane (x_i, x_j), and thus we identify the different two-dimensional matrices of the groups $\boldsymbol{SL}(2, \mathbb{C})$ and the Lorentz group. The representations $\Gamma^{(1/2,0)}$ and $D^{(1/2,0)}$ are, finally, identical and constitute one and the same purely spinorial representation.

The Representation $\Gamma^{0,1/2}$

In accordance with relations (6.1.14) the infinitesimal matrices of the representation $\Lambda^{(0,1/2)}$ are given by

$$\begin{aligned} {B_1}^{(+)} &= \tfrac{1}{2} i\sigma_1, & {B_2}^{(+)} &= -\tfrac{1}{2} i\sigma_2, & {B_3}^{(+)} &= \tfrac{1}{2} i\sigma_3, \\ {B_1}^{(-)} &= \tfrac{1}{2} \sigma_1, & {B_2}^{(-)} &= -\tfrac{1}{2} \sigma_2, & {B_3}^{(-)} &= \tfrac{1}{2} \sigma_3. \end{aligned} \tag{6.3.18}$$

The matrix (6.3.15) of the representation $D^{(0,1/2)}$ allows us to obtain the infinitesimal matrix of this representation which is identical to the matrix ${B_3}^{(+)}$. The representations $D^{(0,1/2)}$ and $\Gamma^{(0,1/2)}$ are identical.

Quite generally, all the representations with $j + j'$ equal to a half-integer are purely spinor representations of the Lorentz group.

6.3.3 The Direct Product of Irreducible Representations

Let us denote by $\overline{D}^{(jj')}$ the irreducible representations of the group of four-dimensional representations $\boldsymbol{SO}(4)$ corresponding to the irreducible representations $D^{(jj')}$ of the proper Lorentz group.

The representations $\overline{D}^{(jj')}$ can be expressed as the direct product of the irreducible representations of two groups of three-dimensional rotations, that is,

$$\overline{D}^{(jj')} = D^{(j)} \otimes D^{(j')}. \tag{6.3.19}$$

Let us look for the decompositions of the direct product of two irreducible representations of $\boldsymbol{SO}(4)$, that is to say,

$$\overline{D}^{(j_1 j'_1)} \otimes \overline{D}^{(j_2 j'_2)} \tag{6.3.20}$$

in irreducible representations. By (6.3.19) we can write

$$\overline{D}^{(j_1 j'_1)} \otimes \overline{D}^{(j_2 j'_2)} = (C^{j_1} \otimes D^{j'_1}) \otimes (D^{j_2} \otimes D^{j'_2}). \tag{6.3.21}$$

We can interchange the order of the factors in the right-hand side of relation (6.3.21), considering the quality of the matrices of a representation, to within a symmetry transformation, whence,

$$\overline{D}^{(j_1 j'_1)} \otimes \overline{D}^{(j_1 j'_2)} = (D^{j_1} \otimes D^{j_2}) \otimes (D^{j'_1} \otimes D^{j'_2}). \tag{6.3.22}$$

Using the Clebsch–Gordan decomposition rule for the direct products of representations of the groups $\boldsymbol{SO}(3)$ we obtain

$$D^{j_1} \otimes D^{j_2} = \sum_{j=|j_1-j_2|}^{j_1+j_2}{}^{\oplus} D_{(j)}, \tag{6.3.23}$$

as well as an analogous expression for $D^{j'_1} \otimes D^{j'_2}$. Substituting these expressions in (6.3.22) we obtain

$$\overline{D}^{(j_1 j'_1)} \otimes \overline{D}^{(j_2 j'_2)} = \sum_{j=|j_1-j_2|}^{j_1+j_2}{}^{\oplus} D^{(j)} \otimes \sum_{j'=|j'_1-j'_2|}^{j'_1+j'_2}{}^{\oplus} D^{(j')}. \tag{6.3.24}$$

Using the new expression (6.3.19) in the opposite sense expression (6.3.24) is written

$$\overline{D}^{(j_1 j'_1)} \otimes \overline{D}^{(j_2 j'_2)} = \sum_{j=|j_1-j_2|}^{j_1+j_2} \sum_{j'=|j'_1-j'_2|}^{j'_1+j'_2}{}^{\oplus} \overline{D}^{(jj')}. \tag{6.3.25}$$

As a result of the relation established between the irreducible representations of the group $\boldsymbol{SO}(4)$ and those of the Lorentz group we obtain a decomposition analogous to (6.3.5) for the direct product of two finite-dimensional irreducible representations of the proper Lorentz group.

6.4 Solved Problems

Exercise **6.1**

To the four-vector $\boldsymbol{X} = (x^0, x^1, x^2, x^3)$ of Minkowski space we make correspond the matrix $\chi = x^i \sigma_i$, where σ_i denotes the Pauli matrices.

1. Write the matrix χ explicitly and calculate the inverse correspondence.

2. Calculate the scalar product $(\boldsymbol{X}, \boldsymbol{X})$ as a function of χ.

3. Let A be a matrix belonging to the group $\boldsymbol{SL}(2, \mathbb{C})$. Determine a transformation which preserves the scalar product of four-vectors.

4. From this deduce that there exists a Lorentz transformation Λ_A corresponding to the matrix A.

5. Show that there exists a homomorphism between the groups $\boldsymbol{SL}(2,\mathbb{C})$ and $\boldsymbol{SO}(3,1)^{\dagger}$.

SOLUTION **6.1**

1. The matrix χ is given by

$$\chi = s^0\sigma_0 + x^1\sigma_1 + x^2\sigma_2 + x^3\sigma_3,$$

that is,

$$\chi = \begin{bmatrix} x^0 + x^3 & x^1 - ix^2 \\ x^1 + ix^2 & x^0 - x^3 \end{bmatrix}.$$

The correspondence is invertible and we obtain

$$x^i = \tfrac{1}{2}\,\mathrm{Tr}\,(\chi\sigma_i).$$

2. The scalar product $(\boldsymbol{X}, \boldsymbol{X})$ is given by

$$(\boldsymbol{X}, \boldsymbol{X}) = (x^0)^2 - (x^1)^2 - (x^2)^2 - (x^3)^2.$$

This is the expression for the determinant of the matrix χ

$$\det\chi = (\boldsymbol{X}, \boldsymbol{X}).$$

3. The matrices of the group $\boldsymbol{SL}(2,\mathbb{C})$ have a determinant which is equal to unity, that is, $\det A = 1$. Let us consider the transformation

$$\chi \to \chi' = A\chi A^{\dagger}.$$

We have

$$\det\chi' = \det(A\chi A^{\dagger}) = \det\chi.$$

Let us bring into correspondence with χ' the four-vector $\boldsymbol{X}'$ such that

$$\chi' = x'^i\sigma_i.$$

We then have

$$\det\chi = (\boldsymbol{X}, \boldsymbol{X}) = \det\chi' = (\boldsymbol{X}', \boldsymbol{X}').$$

4. The Lorentz transformations are linear transformations which leave the scalar product $(\boldsymbol{X}, \boldsymbol{X})$ invariant. Thus there exists a Lorentz transformation Λ_A associated with each matrix A which to each four-vector brings into correspondence the four-vector $\boldsymbol{X}'$ such that

$$\boldsymbol{X}' = \Lambda_A \boldsymbol{X}.$$

5. If we change A into $-A$ we always have the same correspondence

$$\chi \to \chi' = (-A)\chi(-A^\dagger) = A\chi A^\dagger.$$

Therefore we have $\Lambda_A = \Lambda_{-A}$. Thus to a matrix of the group $\boldsymbol{SO}(3,1)^\dagger$ there correspond two matrices A and $-A$ of the group $\boldsymbol{SL}(2,\mathbb{C})$. We therefore have a homomorphism which to an element of $\boldsymbol{SO}(3,1)^\dagger$ brings two elements of $\boldsymbol{SL}(2,\mathbb{C})$ into correspondence. We prove that this homomorphism $\pm A \to \Lambda_A$ is continuous, and that locally we have an isomorphism. The group $\boldsymbol{SL}(2,\mathbb{C})$ constitutes the universal covering group of $\boldsymbol{SO}(3,1)^\dagger$.

EXERCISE **6.2**

The infinitesimal matrices Y_{ij} of the group $\boldsymbol{SO}(3,1)^\dagger$ are given by (6.1.9).

1. Calculate the commutators $[Y_{02}, Y_{01}]$, $[Y_{12}, Y_{01}]$, $[Y_{02}, Y_{12}]$.

2. Compare them with the commutation relations of $\boldsymbol{SO}(3)$.

SOLUTION **6.2**

1. Let us calculate the following matrix products

$$Y_{02}Y_{01} = \begin{bmatrix} 0 & 0 & 1 & 0 \\ 0 & 0 & 0 & 0 \\ 1 & 0 & 0 & 0 \\ 0 & 0 & 0 & 0 \end{bmatrix} \begin{bmatrix} 0 & 1 & 0 & 0 \\ 1 & 0 & 0 & 0 \\ 0 & 0 & 0 & 0 \\ 0 & 0 & 0 & 0 \end{bmatrix} = \begin{bmatrix} 0 & 0 & 0 & 0 \\ 0 & 0 & 0 & 0 \\ 0 & 1 & 0 & 0 \\ 0 & 0 & 0 & 0 \end{bmatrix},$$

$$Y_{01}Y_{02} = \begin{bmatrix} 0 & 1 & 0 & 0 \\ 1 & 0 & 0 & 0 \\ 0 & 0 & 0 & 0 \\ 0 & 0 & 0 & 0 \end{bmatrix} \begin{bmatrix} 0 & 0 & 1 & 0 \\ 0 & 0 & 0 & 0 \\ 1 & 0 & 0 & 0 \\ 0 & 0 & 0 & 0 \end{bmatrix} = \begin{bmatrix} 0 & 0 & 0 & 0 \\ 0 & 0 & 1 & 0 \\ 0 & 0 & 0 & 0 \\ 0 & 0 & 0 & 0 \end{bmatrix}.$$

The commutator $[Y_{02}, Y_{01}] = Y_{02}Y_{01} - Y_{01}Y_{02}$ thus has the expression

$$[Y_{02}, Y_{01}] = Y_{12}.$$

The other commutators are calculated similarly, and we obtain

$$[Y_{12}, Y_{01}] = Y_{02}, \qquad [Y_{02}, Y_{12}] = Y_{01}.$$

2. The matrices Y_{01}, Y_{02}, Y_{03} correspond to the Lorentz transformation which can be represented by 'rotations' in the spacio–temporal planes (x^0, x^1) and (x^0, x^2) as well as in the spacial plane (x^1, x^2). Thus, taking account of the differences in notation, we obtain commutation relations identical to those of the group $\boldsymbol{SO}(3)$.

EXERCISE **6.3**
Consider two spinors (ψ^1, ψ^2) and (ϕ^1, ϕ^2) of four-dimensional space.

1. Show that the product $(\psi^1\phi^2 - \psi^2\phi^1)$ is invariant under all transformations of the group $\boldsymbol{SL}(2, \mathbb{C})$.

2. In order that the product $(\psi^1\phi^2 - \psi^2\phi^1)$ may be written in the classical form $\psi^k\phi_k$ of a scalar product, determine the expression for the 'covariant' components ϕ_k as well as the components of the fundamental tensor.

SOLUTION **6.3**

1. The transformations of the group $\boldsymbol{SL}(2, \mathbb{C})$ are given by (6.2.1) with the condition

$$ad - bc = 1.$$

The transforms of (ψ^1, ψ^2) are

$$\psi'^1 = a\psi^1 + b\psi^2, \qquad \psi'^2 = c\psi^1 + d\psi^2,$$

and similarly for (ϕ^1, ϕ^2), whence

$$\begin{aligned}(\psi'^1\phi'^2 - \psi'^2\phi'^1) &= (ad - bc)\psi^1\phi^2 - (ad - bc)\psi^2\phi^1, \\ &= (\psi^1\phi^2 - \psi^2\phi^1).\end{aligned}$$

2. In order that the scalar product of two spinors may be written in the classical form $\psi^k\phi_k$, where the ψ^k are the 'contravariant' components and the ϕ_k are the 'covariant' components, let us introduce the following covariant components

$$\phi_1 = \phi^2, \qquad \phi_2 = -\phi^1, \qquad \psi_1 = \psi^2, \qquad \psi_2 = -\psi^1.$$

We then obtain

$$\begin{aligned}\psi^1\phi^2 - \psi^2\phi^1 &= \psi^1\phi_1 + \psi^2\phi_2 \\ &= \psi^k\phi_k, \qquad k = 1, 2.\end{aligned}$$

The covariant components are related classically to the contravariant components by

$$\psi_k = g_{ki}\psi^i, \qquad k, i = 1.$$

We thus obtain for the expression for the components of the fundamental tensor

$$g_{11} = 0, \qquad g_{12} = 1, \qquad g_{21} = -1, \qquad g_{22} = 0.$$

This is the fundamental tensor (6.2.7).

EXERCISE **6.4**
Consider two Cartesian frames of reference $Oxyz(t)$ and $O'x'y'z'(t')$ which are moving with respect to each other at a speed v along the axis Ox parallel to $O'x'$.

1. Recall the formulas for the Lorentz transformation between two reference frames. Set

$$\beta = \frac{v}{c}.$$

2. For the moment use the following notations

$$\begin{aligned} ct &= x^0, & x &= x^1, & y &= x^2, & z &= x^3, \\ ct' &= x^{0\prime} & x' &= x^{1\prime}, & y' &= x^{2\prime}, & z' &= x^{3\prime}. \end{aligned} \tag{6.4.26}$$

Set

$$\gamma = (1 - \beta^2)^{-1/2}.$$

Calculate the matrix of the Lorentz transformation between the two reference frames.

3. Show that we can, indeed, set

$$\gamma = \cosh\varphi, \qquad \gamma\beta = \sinh\varphi.$$

Solution **6.4**

1. The Lorentz transformation formulas may be written

$$t' = \frac{t + \dfrac{vx}{c^2}}{(1-\beta^2)^{1/2}},$$

$$x' = \frac{x + vt}{(1-\beta^2)^{1/2}},$$

$$y' = y,$$

$$z' = z.$$

2. The new notations allow us to write the Lorentz transformation formulas in the form

$$x^{0\prime} = \gamma(x^0 + \beta x^1),$$

$$x^{1\prime} = \gamma(\beta x^0 + x^1),$$

$$x^{2\prime} = x^2,$$

$$x^{3\prime} = x^3.$$

Writing these formulas in matrix form we obtain

$$\begin{bmatrix} x^{0\prime} \\ x^{1\prime} \\ x^{2\prime} \\ x^{3\prime} \end{bmatrix} = \begin{bmatrix} \gamma & \gamma\beta & 0 & 0 \\ \gamma\beta & \gamma & 0 & 0 \\ 0 & 0 & 1 & 0 \\ 0 & 0 & 0 & 1 \end{bmatrix} \begin{bmatrix} x^0 \\ x^1 \\ x^2 \\ x^3 \end{bmatrix}.$$

3. In order to justify the changes of variables $\gamma = \cosh\varphi$ and $\gamma\beta = \sinh\varphi$ it is necessary to verify that the new variables satisfy the fundamental relation

$$\cosh^2\varphi - \sinh^2\varphi = 1.$$

This latter equation is certainly satisfied, because we have

$$\gamma^2 - \beta^2\gamma^2 = \frac{1}{1 - \left(\dfrac{v^2}{c^2}\right)} - \frac{v^2}{c^2}\left(\frac{1}{1 - \left(\dfrac{v^2}{c^2}\right)}\right)$$

$$= 1.$$

Respectively replacing γ and $\gamma\beta$ by $\cosh\varphi$ and $\sinh\varphi$ in the matrix obtained in the second question, we obtain the 'rotation' matrix in the plane (x^0, x^1) given by (6.1.4).

7

Dirac Spinors

7.1 The Dirac Equation

In a paper published in 1928 P.A.M. Dirac [8] established a relativistic wave equation for the electron which allows the introduction of the intrinsic magnetic moment of this particle. This equation is invariant under proper Lorentz transformations and the wave function is a spinor. Starting from this article we recover the formation of this equation.

7.1.1 The Classical Relativistic Wave Equation

The classical relativistic Hamiltonism for a point electron moving in an electromagnetic field with a scalar potential A_0 and vector potential $\boldsymbol{A}$ is given by

$$H = \left(\frac{W}{c} + \frac{e}{c}A_0\right)^2 + \left(\boldsymbol{p} + \frac{e}{c}\boldsymbol{A}\right)^2 + m^2c^2, \tag{7.1.1}$$

where $\boldsymbol{p}$ is the momentum vector, W is the kinetic energy of the particle, and m its rest mass. A relativistic wave equation for quantum theory can be obtained by carrying out in (7.1.1) the same substitutions as in the nonrelativistic theory, that is to say,

$$W = i\hbar\frac{\partial}{\partial t}, \qquad P_j = -i\hbar\frac{\partial}{\partial x_j}, \qquad j = 1, 2, 3. \tag{7.1.2}$$

Thus we obtain the relativistic wave equation

$$\left[\left(\frac{i\hbar}{c}\frac{\partial}{\partial t}+\frac{e}{c}A_0\right)^2+\sum_j\left(-i\hbar\frac{\partial}{\partial x_j}+\frac{e}{c}A_j\right)^2+m^2c^2\right]\Psi=0. \quad (7.1.3)$$

On examining certain results of (7.1.3) Dirac was led to the conclusion that the quantum theory equation would have to have only first order partial derivatives with respect to the variables x, y, z, t.

7.1.2 The Dirac Equation for a Free Particle

For a free particle first of all, let us seek a linear wave which is first order. Let us write

$$P_0 = i\hbar\frac{\partial}{\partial t}. \quad (7.1.4)$$

Such an equation must be invariant under a Lorentz transformation, and is sought in the form

$$(P_0+\alpha_1 P_1+\alpha_2 P_2+\alpha_3 P_3+\beta)\Psi=0, \quad (7.1.5)$$

where the operators $\alpha_1, \alpha_2, \alpha_3, \beta$ are to be determined. Since we consider, for the moment, empty space, all the points of this space are equivalent, and the Hamiltonian must not depend on the variables t, x_1, x_2, x_3. This means that $\alpha_1, \alpha_2, \alpha_3, \beta$ are independent of these variables and therefore commute with the operators P_j, $j = 1, \ldots, 4$.

Equation (7.1.5) must allow us to recover the relativistic wave equation (7.1.3), which in the present case reduces to

$$(-{P_0}^2+\boldsymbol{P}^2+m^2c^2)\Psi=0, \quad (7.1.6)$$

where $\boldsymbol{P}$ is the vector operator which has P_1, P_2, P_3 as its components. Multiplying (7.1.5) by the operator $(-P_0+\alpha_1 P_1+\alpha_2 P_2+\alpha_3 P_3+\beta)$ we obtain

$$(-P_0+\alpha_1 P_1+\alpha_2 P_2+\alpha_3 P_3+\beta\times(P_0+\alpha_1 P_1+\alpha_2 P_2+\alpha_3 P_3+\beta)=0. \quad (7.1.7)$$

Let us expand the latter relation, which gives us

$$\Big[-{P_0}^2+\sum {\alpha_1}^2{P_1}^2+\sum(\alpha_1\alpha_2+\alpha_2\alpha_1)P_1P_2$$
$$+\beta^2+\sum(\alpha_1\beta+\beta\alpha_1)P_1\Big]\Psi=0, \quad (7.1.8)$$

where the summation symbols $\sum$ mean that we must carry out a cyclic permutation of the indices 1, 2, 3. In order to identify (7.1.6) and (7.1.8) it is necessary to have

$$\begin{aligned} {\alpha_j}^2 &= 1, & \alpha_j\alpha_k + \alpha_k\alpha_j &= 0, & j &\neq k, \\ \beta^2 &= m^2c^2, & \alpha_j\beta + \beta\alpha_j &= 0, \end{aligned} \tag{7.1.9}$$

with $j, k = 1, 2, 3$. Let us set $\beta = mc\alpha_4$, then conditions (7.1.9) may be written

$$\alpha_j\alpha_k + \alpha_k\alpha_j = 2\delta_{jk}, \qquad j, k = 1, 2, 3, 4. \tag{7.1.10}$$

Let us look for matrices representing the operators α_j which satisfy relations (7.1.10). Dirac proposed using the following matrices, which are formed with the help of the Pauli matrices denoted σ_j

$$\Sigma_j = \begin{bmatrix} \sigma_j & 0 \\ 0 & \sigma_j \end{bmatrix}, \qquad \rho_1 = \begin{bmatrix} 0 & \sigma_1 \\ \sigma_1 & 0 \end{bmatrix}, \qquad \rho_3 = \begin{bmatrix} \mathbb{1} & 0 \\ 0 & -\mathbb{1} \end{bmatrix}, \tag{7.1.11}$$

where $\mathbb{1}$ is the unit matrix of order 2. The matrices α_j are then defined from the preceding matrices, that is to say,

$$\alpha_j = \rho_1\Sigma_j, \qquad j = 1, 2, 3, \qquad \alpha_4 = \psi_3. \tag{7.1.12}$$

It is easily verified that all conditions (7.1.10) are satisfied. Equation (7.1.5) then takes the form

$$(P_0 + \rho_1(\mathbf{\Sigma}\cdot\boldsymbol{P}) + \rho_3 mc)\Psi = 0, \tag{7.1.13}$$

where $\mathbf{\Sigma}$ represents the vector having the matrices $\Sigma_1, \Sigma_2, \Sigma_3$ as its components. Equation (7.1.13) is the DIRAC EQUATION for a free particle.

7.1.3 A Particle in an Electromagnetic Field

The wave equation for an electron in an electromagnetic field with scalar potential A_0 and vector potential $\boldsymbol{A}$ is then obtained by carrying out the classical substitutions of nonrelativistic quantum mechanics. Replacing P_0 by $P_0 + (e/c)A_0$ and $\boldsymbol{P}$ by $\boldsymbol{P} + (e/c)\boldsymbol{A}$ in the Hamiltonian for no field, from (7.1.13) we obtain

$$\left\{P_0 + \frac{e}{c}A_0 + \rho_1\left[\mathbf{\Sigma}\cdot\left(\boldsymbol{P} + \frac{e}{c}\boldsymbol{A}\right)\right] + \rho_3 mc\right\}\Psi = 0. \tag{7.1.14}$$

The matrices of order 4 which appear in (7.1.14) require that Ψ has four components, denoted ψ_j, $j = 1, \ldots, 4$. Let us write

$$P_0{}' = P_0 + \frac{e}{c}A_0 \qquad \text{and} \qquad P_j{}' = P_j + \frac{e}{c}A_j.$$

The expansion of these matrices appearing in (7.1.14) then gives us four equations

$$\begin{aligned}
(P_0' + mc)\psi_1 + (P_1' + iP_2')\psi_4 + P_3'\psi_3 &= 0, \\
(P_0' + mc)\psi_2 + (P_1' - iP_2')\psi_3 - P_3'\psi_4 &= 0, \\
(P_0' - mc)\psi_3 + (P_1' + iP_2')\psi_2 + P_3'\psi_1 &= 0, \\
(P_0' - mc)\psi_4 + (P_1' - iP_2')\psi_1 - P_3'\psi_2 &= 0,
\end{aligned} \tag{7.1.15}$$

Combined with each other, these equations allow us to recover the second-order equation (7.1.3) which the ψ_k satisfy. We obtain an obvious result since the matrices α_j satisfy exactly the compatibility relations (7.1.10). Furthermore, these equations introduce the exact expression for the intrinsic magnetic moment of the electron into the relativistic equation which Ψ satisfies.

7.2 Relativistic Invariance of the Dirac Equation

7.2.1 The Relativistic Invariance Condition

In his paper previously cited, Dirac proved that (7.1.14) is invariant under Lorentz transformations. Conversely, we can find matrices α_j defined by (7.1.12) from the relativistic invariance of an equation in first-order partial derivatives, and which can be written in the general form

$$(\alpha_0 P_0 + \alpha_1 P_1 + \alpha_2 P_2 + \alpha_3 P_3 + \beta)\Psi = 0. \tag{7.2.1}$$

Let us study the invariance of this equation under a Lorentz transformation transforming a spacio–temporal point defined by the position vector $\boldsymbol{r}$ into another point $\boldsymbol{r}'$, such that

$$\boldsymbol{r}' = \Lambda \boldsymbol{r}. \tag{7.2.2}$$

Under such a transformation the wave function is transformed according to the relation

$$\Psi'(\boldsymbol{r}') = M(\Lambda)\Psi(\boldsymbol{r}). \tag{7.2.3}$$

The matrices $M(\Lambda)$ form a representation of the Lorentz group. On the other hand, let us write (7.2.1) in the new frame of reference. To do so, by applying the inverse transformation we obtain

$$\Psi(\boldsymbol{r}) = M^{-1}(\Lambda)\Psi'(\boldsymbol{r}'). \tag{7.2.4}$$

Let us write the elements of the matrices of Λ as Λ_{ij}. Relation (7.2.2) gives us

$$x_i{}' = \Lambda_{ij} x_j, \tag{7.2.5}$$

where the sum is carried out over $j = 1, 2, 3$ with Einstein's convention. As a consequence we have

$$\frac{\partial}{\partial x_i} = \Lambda_{ji} \frac{\partial}{\partial x_j{}'} . \tag{7.2.6}$$

Substituting (7.2.3) and (7.2.6) into (7.2.1) we obtain

$$\alpha_i M^{-1} \left(i\hbar \frac{\partial \Psi'}{\partial x_j{}'} \right) \Lambda_{ij} + \beta M^{-1} \Psi' = 0. \tag{7.2.7}$$

Let us assume that the matrix β is invertible. Without affecting the generality of the argument we can also assume that β is a matrix which is a multiple of the unit matrix. In fact, we can always multiply both sides of (7.2.7) by β^{-1}. Multiplying (7.2.7) on the left by the matrix $M(\Lambda)$ we obtain

$$\Lambda_{ji} M \alpha_i M^{-1} \left(i\hbar \frac{\partial \Psi'}{\partial x_j{}'} \right) + \beta \Psi' = 0. \tag{7.2.8}$$

In order that (7.2.1) be invariant with respect to Lorentz transformations it is necessary that it be of the same form as (7.2.8), whence,

$$\alpha_j = \Lambda_{ji} M \alpha_i M^{-1}. \tag{7.2.9}$$

This latter relation constitutes the condition of invariance of the system of partial differential equations (7.2.1), and this condition allows us to determine the matrices α_j explicitly. We then obtain the matrices defined by (7.1.12).

From equality (7.2.9), on the other hand, we obtain

$$M^{-1} \alpha_j M = \Lambda_{ji} \alpha_i. \tag{7.2.10}$$

If we consider the vector α as having the matrices α_j as its components, the preceding relation shows that the vector α transforms as a four-dimensional vector.

7.2.2 The Type of Representation for the Wave Function

By hypothesis the function Ψ of the relativistic equation (7.2.1) transforms according to some representation of the Lorentz group. Let us determine the type of irreducible representation, of minimal dimension, of the Lorentz group which satisfies (7.2.1). To do so let us go to the term $\beta\Psi$ in the right-hand side. We obtain

$$(\alpha_0 P_0 + \alpha_1 P_1 + \alpha_2 P_2 + \alpha_3 P_3)\Psi = -\beta\Psi. \tag{7.2.11}$$

The quantity appearing on the right-hand side transforms under a certain irreducible representation $D^{(jj')}$. Against this, in the left-hand side there appears the four-dimensional vector $\boldsymbol{P}' = (P_0, P_1, P_2, P_3)$ which transforms under the irreducible representation $D^{(1/2,1/2)}$. The quantity appearing in the left-hand side thus transforms under the representation $D^{(1/2,1/2)} \otimes D^{(jj')}$.

If we assume, starting from the representation of smallest dimension, that the right-hand side of (7.2.11) transforms according to the one-dimensional representation of the identity $D^{(0,0)}$, the left hand side transforms according to the representation $D^{(1/2,1/2)} \otimes D^{(0,0)}$. Now, equality of two quantities transforming according to two different representations is impossible.

Let us consider, for the moment, a wave function transforming under the representation $D^{(1/2,0)}$. The left hand term would then transform under the representation $D^{(1/2,1/2)} \otimes D^{(1/2,0)}$ which decomposes, according to formula (6.2.25), as

$$D^{(1/2,1/2)} \otimes D^{(1/2,0)} = D^{(0,1/2)} \oplus D^{(1,1/2)}.$$

As the two sides of (7.2.11) do not transform under the same representation, the hypothesis must therefore be rejected. Finally, we will obtain a coherent irreducible representation if we assume that the wave function transforms under the representation

$$D^{(1/2,0)} \oplus D^{(0,1/2)}. \tag{7.2.12}$$

This means that the first two components ψ_1, ψ_2 of the wave function Ψ transform under the representation $D^{(1/2,0)}$, and the two others, Ψ_3 and ψ_4, under the representation $D^{(0,1/2)}$. The wave function Ψ is a four-component spinor which transforms in a particular way. This type of spinor is called a *bispinor*.

7.2.3 The Link Between a Spinor and a Four-Vector

In order to establish the Dirac equation directly in the spinor representation let us study the relations between a spinor and a four-vector. We have already seen in the course of Chapter 6 the equivalence between the four-dimensional representations obtained from spinors and four-vectors. Let us return in another form to the study of these relations.

A spinor $\psi^{\alpha\dot{\beta}}$ with one undotted and one dotted index possesses four independent components, similarly to a four-vector. Both can therefore allow us a priori to obtain the same representation of the Lorentz group. Thus there exists one relation between the components of the four-vectors and spinors with four components.

In order to establish this relation let us use the property that spinors in three-dimensional space and those of four-dimensional space have the same behavior with respect to purely spacial rotations. For a spinor of three-dimensional space we have the correspondence formulas given by (1.1.24). Let us write A_x, A_y, A_y for the components of a vector $\boldsymbol{A}$ and write

$$\psi^{11} = (\phi)^2, \qquad \psi^{12} = \psi\phi, \qquad \psi^{22} = (\psi)^2$$

for the components of a spinor of three-dimensional space. Relations (1.1.24) may then be written in the form

$$\begin{aligned} A_x &= \tfrac{1}{2}(\psi^{22} - \psi^{11}) = \tfrac{1}{2}({\psi^2}_1 + {\psi^1}_2), \\ A_y &= -\tfrac{1}{2}i(\psi^{22} + \psi^{11}) = \tfrac{1}{2}i({\psi^1}_2 - {\psi^2}_1), \\ A_z &= \tfrac{1}{2}(\psi^{12} + \psi^{21}) = \tfrac{1}{2}({\psi^1}_1 - {\psi^2}_2). \end{aligned} \tag{7.2.13}$$

Passing to the four-dimensional case we shall replace the components ${\psi^\alpha}_\beta$ with $\zeta^{\alpha\dot{\beta}}$ and we shall consider A_x, A_y, A_z to be the contravariant components A^1, A^2, A^3 of a four-vector.

As for the expression for the time component A^0, it results from the particle density in relativity theory being no longer a scalar but the time component of a four-vector. This particle density is proportional to the probability density, and thus A^0 must be transformed as the latter. Therefore we have

$$A^0 \sim \zeta^{i\dot{1}} + \zeta^{2\dot{2}}.$$

The coefficient of proportionality is determined in such a way that the scalar $\alpha_{\alpha\dot{\beta}}\zeta^{\alpha\dot{\beta}}$ coincides with the scalar

$$2A_\nu A^\nu = 2\boldsymbol{A}^2.$$

Thus we obtain the correspondence relations sought

$$\begin{aligned} A^1 &= \tfrac{1}{2}(\zeta^{1\dot{2}} + \zeta^{2\dot{1}}), \qquad & A^2 &= \tfrac{1}{2}i(\zeta^{1\dot{2}} - \zeta^{2\dot{1}}), \\ A^3 &= \tfrac{1}{2}(\zeta^{i\dot{1}} - \zeta^{2\dot{2}}), \qquad & A^0 &= \tfrac{1}{2}(\zeta^{1\dot{1}} + \zeta^{2\dot{2}}). \end{aligned} \tag{7.2.14}$$

Conversely, we obtain the formulas defining the components of a four-spinor as a function of those of a four-vector

$$\begin{aligned} \zeta^{1\dot{1}} &= \zeta_{2\dot{2}} = A^3 + A^0, \qquad & \zeta^{2\dot{2}} &= \zeta_{1\dot{1}} = A^0 - A^3, \\ \zeta^{1\dot{2}} &= -\zeta_{2\dot{1}} = A^1 - iA^2, \qquad & \zeta^{2\dot{1}} &= -\zeta_{1\dot{2}} = A^1 + iA^2. \end{aligned} \tag{7.2.15}$$

It is verified that we obtain the scalar

$$\zeta_{\alpha\dot{\beta}}\zeta^{\alpha\dot{\beta}} = 2\boldsymbol{A}^2. \tag{7.2.16}$$

We also obtain the relation

$$\zeta_{\alpha\dot{\beta}}\zeta^{\nu\dot{\beta}} = \delta_{\alpha}{}^{\nu}\boldsymbol{A}^2. \tag{7.2.17}$$

Let us write ζ symbolically for the matrix having as its elements $\zeta^{\alpha\dot{\beta}}$ with upper indices. Formulas (7.2.15) may then be written in the matrix form

$$\zeta = \boldsymbol{A}\cdot\boldsymbol{\sigma} + A^0\mathbb{1}, \tag{7.2.18}$$

where $\boldsymbol{\sigma}$ is the vector having as its components the Pauli matrices, and $\mathbb{1}$ is the unit matrix of order 2.

7.2.4 Dirac's Equation in the Spinor Representation

Let us write P_0, P_x, P_y, P_z for the components of the four-momentum operator. The latter is a four-vector to which there corresponds a spinor operator $P_{\alpha\beta}$ defined with the help of relations (7.2.15), that is,

$$\begin{aligned} P^{1\dot{1}} &= P_{2\dot{2}} = P_z + P_0, & P^{2\dot{2}} &= P_{1\dot{1}} = P_0 - P_z, \\ P^{1\dot{2}} &= -P_{2\dot{1}} = P_x - iP_y, & P^{2\dot{1}} &= -P_{1\dot{2}} = P_x + iP_y. \end{aligned} \tag{7.2.19}$$

The description of a particle with spin $\frac{1}{2}$ in an arbitrary reference frame introduces two types of four-spinor, dotted and undotted spinors. We shall write them ζ^{α} and $\eta_{\dot{\beta}}$.

The wave equation written in spinor form is a relation in partial derivatives between the components of the spinors realized by starting from the four-momentum operator $P_{\alpha\dot{\beta}}$. The condition for relativistic invariance leads to the following system of equations

$$P^{\alpha\dot{\beta}}\eta_{\dot{\beta}} = m\zeta^{\alpha}, \qquad P_{\alpha\dot{\beta}}\zeta^{\alpha} = m\eta_{\dot{\beta}}, \tag{7.2.20}$$

where m is a dimensional constant. It is not useful to introduce two different constants m_1 and m_2 for each of these equations, since a change of a spinor by a given constant allows us to reduce to (7.2.20).

Let us eliminate the spinor $\eta_{\dot{\beta}}$, for example, from (7.2.20) by substituting the former from the second equation into the first. We obtain

$$P^{\alpha\dot{\beta}}\eta_{\dot{\beta}} = \frac{1}{m}P^{\alpha\dot{\beta}}P_{\nu\dot{\beta}}\zeta^{\nu} = m\zeta^{\alpha}. \tag{7.2.21}$$

Formula (7.2.17) gives

$$P^{\alpha\dot{\beta}}P_{\nu\dot{\beta}} = \delta^{\alpha}{}_{\nu}\boldsymbol{P}^2,$$

and (7.2.21) may be written

$$(\boldsymbol{P}^2 - m^2)\zeta^{\nu} = 0. \tag{7.2.22}$$

This equation shows that the constant m is the mass of the particle. The relativistic wave equation represented by the system of (7.2.20), where m is the mass of the particle, is the Dirac equation in spinor form.

Let us write (7.2.20) in a more classical symbolic form. Let us write as P the matrix having as its matrix elements $P^{\alpha\dot{\beta}}$ with upper indices. Formulas (7.2.19) may then be written in the matrix form

$$P = \boldsymbol{P}\cdot\boldsymbol{\sigma} + P_0 \mathbb{1}. \tag{7.2.23}$$

It is a form analogous to relation (7.2.18). The transcription of (7.2.20) then gives us

$$(P_0 + \boldsymbol{P}\cdot\boldsymbol{\sigma})\eta = m\zeta, \qquad (P_0 - \boldsymbol{P}\cdot\boldsymbol{\sigma})\zeta = m\eta, \tag{7.2.24}$$

where the symbols ζ and η represent, for the moment, two-component spinors

$$\zeta = \begin{bmatrix} \zeta^1 \\ \zeta^2 \end{bmatrix}, \qquad \eta = \begin{bmatrix} \eta_{\dot{1}} \\ \eta_{\dot{2}} \end{bmatrix}. \tag{7.2.25}$$

The spinor ζ transforms according to the irreducible representation $D^{(1/2,0)}$ and the spinor η according to the representation $D^{(0,1/2)}$. We recover the irreducible representation $D^{(1/2,0)} \oplus D^{(0,1/2)}$ of the Dirac bispinors Ψ with four components.

7.2.5 The Symmetric Form of the Dirac Equation

Dirac's equation can be put into different forms. In particular, let us write the system of equations (7.2.24) in the matrix form

$$\begin{bmatrix} 0 & P_0 + \boldsymbol{P}\cdot\boldsymbol{\sigma} \\ P_0 - \boldsymbol{P}\cdot\boldsymbol{\sigma} & 0 \end{bmatrix} \begin{bmatrix} \zeta \\ \eta \end{bmatrix} = m \begin{bmatrix} \zeta \\ \eta \end{bmatrix}. \tag{7.2.26}$$

Let us introduce the following notations

$$\gamma^0 = \begin{bmatrix} 0 & \mathbb{1} \\ \mathbb{1} & 0 \end{bmatrix} \qquad \gamma^k = \begin{bmatrix} 0 & -\sigma_k \\ \sigma_k & 0 \end{bmatrix}, \qquad \begin{bmatrix} \zeta \\ \eta \end{bmatrix}, \tag{7.2.27}$$

where the σ_k are the Pauli matrices. Let us write as $\boldsymbol{\gamma}$ the vector with components $\gamma^1, \gamma^2, \gamma^3$. The latter are called the DIRAC MATRICES. These notations allow us to write (7.2.26) in the form

$$(P_0\gamma^0 - \boldsymbol{P}\cdot\boldsymbol{\gamma})\Psi = m\Psi. \tag{7.2.28}$$

The Standard Representation

When the velocities become small the particle must be described by a single two-component spinor, as is the case in the nonrelativistic theory.

If we pass to the limit $\boldsymbol{P} \to 0$, $W \to m$ in (7.2.24) this yields $\zeta = \eta$ and the two spinors forming the bispinor coincide. However, although only two components of Ψ are independent for small velocities, all the four remain nonzero if we use the spinor representation (7.2.24). A more useful representation in this case is that in which in the limit two of the components of Ψ vanish.

For that to be so, instead of ζ and η let us introduce the following linear combinations

$$\phi = \frac{1}{\sqrt{2}}(\zeta + \eta), \qquad \chi = \frac{1}{\sqrt{2}}(\zeta - \eta). \tag{7.2.29}$$

Then for a particle at rest $\chi = 0$. The equations for ϕ and χ are obtained easily from (7.2.24) by adding and subtracting them, that is,

$$P_0\phi - \boldsymbol{P}\cdot\boldsymbol{\sigma}\chi = m\phi, \qquad -P_0\chi + \boldsymbol{P}\cdot\boldsymbol{\sigma}\phi = m\chi. \tag{7.2.30}$$

This representation of the bispinor Ψ is called the *standard representation* of Ψ.

7.3 Solved Problems

Exercise **7.1**

Let $\boldsymbol{\theta} = \boldsymbol{n}_1\theta$ and $\boldsymbol{\varphi} = \boldsymbol{n}_2\varphi$ respectively be the parameters of a rotation and of a pure Lorentz transformation. As the Lorentz group can be essentially identified with the group $\boldsymbol{SU}(2)\otimes\boldsymbol{SU}(2)$ we can write the following matrix representations of the Lorentz group

— for the representation $D^{(1/2,0)}$: $M = \exp[\frac{1}{2}i\boldsymbol{\sigma}\cdot(\boldsymbol{\theta} - i\boldsymbol{\varphi})]$;

— for the representation $D^{(0,1/2)}$: $N = \exp[\frac{1}{2}i\boldsymbol{\sigma}\cdot(\boldsymbol{\theta} + i\boldsymbol{\varphi})]$.

Show that the matrices are connected by the relations

$$N = (-i\sigma_2)M^*(-i\sigma_2)^{-1}.$$

Solution **7.1**

We note that the following matrix relations hold

$$\begin{aligned} \sigma_2\boldsymbol{\sigma}^*\sigma_2 &= (-\sigma_2)^2\boldsymbol{\sigma} \\ &= -\boldsymbol{\sigma}. \end{aligned}$$

Taking these relations into account, the expression for N becomes

$$\begin{aligned}(-i\sigma_2)M^*(-i\sigma_2)^{-1} &= (-i\sigma_2)\exp[-\tfrac{1}{2}i\boldsymbol{\sigma}^*\boldsymbol{\cdot}(\boldsymbol{\theta}+i\boldsymbol{\varphi})](-i\sigma_2)\\ &= (\sigma_2)^2\exp[\tfrac{1}{2}i\boldsymbol{\sigma}\boldsymbol{\cdot}(\boldsymbol{\theta}+i\boldsymbol{\varphi})]\\ &= N.\end{aligned}$$

The representations $D^{(1/2,0)}$ and $D^{(0,1/2)}$ are not equivalent.

EXERCISE **7.2**
Let Φ_{R} and Φ_{L} be the spinors corresponding respectively to the representations $D^{(1/2,0)}$ and $D^{(0,1/2)}$.

1. Write the transformations of these spinors for the case of a pure Lorentz transformation, using the matrices given in Exercise 7.1.

2. Let $\boldsymbol{\varphi} = \boldsymbol{n}\varphi$ be the parameter characterizing a pure Lorentz transformation. By Exercise 6.4 we have the relations

$$\begin{aligned}\gamma &= \left(1-\frac{v^2}{c^2}\right)^{-1/2} = \cosh\varphi,\\ \beta &= \frac{v}{c},\\ \gamma\beta &= \sinh\varphi.\end{aligned}$$

   Write the equations obtained in the first question in terms of $\boldsymbol{n}$, φ, and γ.

3. Assume that the spinors Φ_{R} and Φ_{L} relate to a particle at rest, and write the corresponding spinors as $\Phi_{\mathrm{R}}(0)$ and $\Phi_{\mathrm{L}}(0)$. This particle acquires a momentum $\boldsymbol{p}$ under the action of a pure Lorentz transformation. We write $\Phi_{\mathrm{R}}(\boldsymbol{p})$ and $\Phi_{\mathrm{L}}(\boldsymbol{p})$ for the spinors of the particle in motion. Let E be the total energy of the particle and let m be its mass. Write the transformation equations obtained in the second question in terms of E, m, and $\boldsymbol{p}$, taking $c = 1$.

4. When a particle is at rest it must be described by a single two-component spinor which is a solution of Pauli's equation. We then have $\Phi_{\mathrm{R}}(0) = \Phi_{\mathrm{L}}(0)$. Prove that there hold the relations

$$\begin{aligned}-m\Phi_{\mathrm{R}}(\boldsymbol{p}) + (P_0 + \boldsymbol{\sigma}\boldsymbol{\cdot}\boldsymbol{P})\Phi_{\mathrm{L}}(\boldsymbol{p}) &= 0,\\ (P_0 - \boldsymbol{\sigma}\boldsymbol{\cdot}\boldsymbol{P})\Phi_{\mathrm{R}}(\boldsymbol{p}) - m\Phi_{\mathrm{L}}(\boldsymbol{p}) &= 0,\end{aligned}$$

   where $P_0 = E$ and $\boldsymbol{P}$ is the momentum operator.

5. Define the spinor

$$\Psi = \begin{bmatrix} \Phi_{\mathrm{R}}(\boldsymbol{p}) \\ \Phi_{\mathrm{L}}(\boldsymbol{p}) \end{bmatrix}.$$

Recover the Dirac equation for the spinor Ψ.

SOLUTION **7.2**

1. In the case of a pure Lorentz transformation we have $\boldsymbol{\theta} = 0$. In this case the spinors of the representations $D^{(1/2,0)}$ and $D^{(0,1/2)}$ transform according to the relations

$$\begin{aligned} \Phi_{\mathrm{R}}{}' &= \exp(\tfrac{1}{2}\boldsymbol{\sigma}\cdot\boldsymbol{\varphi})\Phi_{\mathrm{R}}, \\ \Phi_{\mathrm{L}}{}' &= \exp(-\tfrac{1}{2}\boldsymbol{\sigma}\cdot\boldsymbol{\varphi})\Phi_{\mathrm{L}}. \end{aligned}$$

2. The equations obtained in the first question may be written

$$\begin{aligned} \Phi_{\mathrm{R}}{}' &= (\cosh\tfrac{1}{2}\varphi + \boldsymbol{\sigma}\cdot\boldsymbol{n}\sinh\tfrac{1}{2}\varphi)\Phi_{\mathrm{R}}, \\ \Phi_{\mathrm{L}}{}' &= (\cosh\tfrac{1}{2}\varphi - \boldsymbol{\sigma}\cdot\boldsymbol{n}\sinh\tfrac{1}{2}\varphi)\Phi_{\mathrm{L}}. \end{aligned}$$

On the other hand, the hyperbolic trigonometric relations give

$$\begin{aligned} \cosh\tfrac{1}{2}\varphi &= \left(\frac{\gamma+1}{2}\right)^{1/2}, \\ \sinh\tfrac{1}{2}\varphi &= \left(\frac{\gamma-1}{2}\right)^{1/2}. \end{aligned}$$

The preceding equations can then be written in terms of γ

$$\begin{aligned} \Phi_{\mathrm{R}}{}' &= \left(\left(\frac{\gamma+1}{2}\right)^{1/2} + \boldsymbol{\sigma}\cdot\boldsymbol{n}\left(\frac{\gamma-1}{2}\right)^{1/2}\right)\Phi_{\mathrm{R}}. \\ \Phi_{\mathrm{L}}{}' &= \left(\left(\frac{\gamma+1}{2}\right)^{1/2} - \boldsymbol{\sigma}\cdot\boldsymbol{n}\left(\frac{\gamma-1}{2}\right)^{1/2}\right)\Phi_{\mathrm{L}}. \end{aligned}$$

3. A particle moving freely has the relativistic energy

$$E = mc^2\gamma.$$

If we set $c = 1$ then $\gamma = E/m$. The transformation equations obtained in the second question may then be written

$$\begin{aligned} \Phi_{\mathrm{R}}(\boldsymbol{p}) &= \frac{1}{(2m(E+m))^{1/2}}(E + m + \boldsymbol{\sigma}\cdot\boldsymbol{p})\Phi_{\mathrm{R}}(0), \\ \Phi_{\mathrm{L}}(\boldsymbol{p}) &= \frac{1}{(2m(E+m))^{1/2}}(E + m - \boldsymbol{\sigma}\cdot\boldsymbol{p})\Phi_{\mathrm{L}}(0). \end{aligned}$$

4. Taking $\Phi_{\rm R}(0) = \Phi_{\rm L}(0)$ into account the preceding equations become, by combining them,

$$\begin{aligned} \Phi_{\rm R}(\boldsymbol{p}) &= \frac{1}{m}(E + \boldsymbol{\sigma}\cdot\boldsymbol{p})\Phi_{\rm L}(\boldsymbol{p}), \\ \Phi_{\rm L}(\boldsymbol{p}) &= \frac{1}{m}(E - \boldsymbol{\sigma}\cdot\boldsymbol{p})\Phi_{\rm R}(\boldsymbol{p}). \end{aligned}$$

Set $E = P_0$ and replace $\boldsymbol{p}$ by the operator $\boldsymbol{P}$. Regrouping the spinors to the left-hand side we obtain the equations required

$$\begin{aligned} -m\Phi_{\rm R}(\boldsymbol{p}) + (P_0 + \boldsymbol{\sigma}\cdot\boldsymbol{P})\Phi_{\rm L}(\boldsymbol{p}) &= 0, \\ (P_0 - \boldsymbol{\sigma}\cdot\boldsymbol{P})\Phi_{\rm R}(\boldsymbol{p}) - m\Phi_{\rm L}(\boldsymbol{p}) &= 0. \end{aligned}$$

5. Write the preceding equation in matrix form, which yields

$$\begin{bmatrix} -m & P_0 + \boldsymbol{\sigma}\cdot\boldsymbol{P} \\ P_0 - \boldsymbol{\sigma}\cdot\boldsymbol{P} & -m \end{bmatrix} \begin{bmatrix} \Phi_{\rm R}(\boldsymbol{p}) \\ \Phi_{\rm L}(\boldsymbol{p}) \end{bmatrix} = 0. \tag{7.3.1}$$

This is Dirac's equation given in (7.3.1).

8

Clifford and Lie Algebras

8.1 Lie Algebras

In this chapter we give some definitions and theorems that will enable us to put the results obtained in the course of the previous chapter in a broader context. Thus the infinitesimal method which we have used for different groups, consisting in calculating the infinitesimal generators, was developed by Sophus Lie. It has the effect of reducing the study of Lie groups to the study of what are called their *Lie algebras*.

8.1.1 The Definition of an Algebra

An ALGEBRA is a vector space V provided with an internal composition law called the PRODUCT. Thus with every pair of vectors $(\boldsymbol{x}, \boldsymbol{y})$ of V we associate a vector $\boldsymbol{z}$ of V denoted

$$\boldsymbol{z} = \boldsymbol{x} * \boldsymbol{y}. \tag{8.1.1}$$

The vector $\boldsymbol{z}$ is called to the product of $\boldsymbol{x}$ with $\boldsymbol{y}$. This product must be bilinear, that is,

$$\left(\sum_i \alpha_i \boldsymbol{x}_i\right) * \left(\sum_j \beta_j \boldsymbol{y}_j\right) = \sum_i \sum_j \alpha_i \beta_j (\boldsymbol{x}_i * \boldsymbol{y}_j). \tag{8.1.2}$$

The algebra is called associative if

$$\boldsymbol{x} * (\boldsymbol{y} * \boldsymbol{z}) = (\boldsymbol{x} * \boldsymbol{y}) * \boldsymbol{z}. \tag{8.1.3}$$

Examples

The square matrices of order n form an n^2-dimensional vector space V_n. The classical matrix product provides V_n with an associative algebra structure.

Another example is that of the vector space formed by the vectors of $\mathbb{R}^3$ provided with the classical vector space product $\boldsymbol{z} = (\boldsymbol{x} \times \boldsymbol{y})$. The latter certainly is a bilinear product and the vector space $\mathbb{R}^3$ is then an algebra.

Let us also consider the infinitesimal matrices L_1, L_2, L_3 of the group $\boldsymbol{SO}(3)$. These matrices are able to serve as a basis of a three-dimensional vector space V_3. Let us define the following product

$$\begin{aligned} L_i * L_j &= L_i L_j - L_j L_i \\ &= [L_i, L_j], \end{aligned} \tag{8.1.4}$$

which is called the COMMUTATOR of L_i and L_j. Such a product again gives an element of V_3, since we have for the commutator of these matrices

$$L_i L_j - L_j L_i = \varepsilon_{ijk} L_k. \tag{8.1.5}$$

It is easily verified that the commutator is a bilinear operation and the vector space V_3 provided with the product (8.1.4) is then a three-dimensional algebra.

8.1.2 Lie Algebras

EXAMPLE **8.1.** Let us consider the algebra V_3 which has just been defined and has the infinitesimal matrices L_1, L_2, L_3 as basis matrices. The product (8.1.4) satisfies the obvious property

$$[L_i, L_j] = -[L_j, L_i]. \tag{8.1.6}$$

On the other hand, using the value of the commutator (8.1.5) we can verify that the following matrix relation holds

$$[L_i, [L_j, L_k]] + [L_j, [L_k, L_i]] + [L_k, [L_i, l_j]] = 0. \tag{8.1.7}$$

The algebra V_3 which thus satisfies properties (8.1.6) and (8.1.7) is an example of a LIE ALGEBRA. This algebra, defined by the infinitesimal matrices of the group $\boldsymbol{SO}(3)$ is denoted $\mathfrak{so}(3)$. ■

DEFINITION **8.1.** *A Lie algebra $\mathfrak{g}$ is a vector space defined over a field K, and in which a product denoted $[\cdot, \cdot]$ called the Lie bracket is defined, and which has the following properties:*

1. *If $\boldsymbol{x}$ and $\boldsymbol{y}$ are elements of $\mathfrak{g}$ then $\boldsymbol{z} = [\boldsymbol{x}, \boldsymbol{y}]$ is an element of $\mathfrak{g}$.*
2. *$[\boldsymbol{x}, \alpha\boldsymbol{y} + \beta\boldsymbol{z}] = \alpha[\boldsymbol{x}, \boldsymbol{y}] + \beta[\boldsymbol{x}, \boldsymbol{z}]$, where α and β are elements of K.*

3. The product is anticommutative

$$[\boldsymbol{x}, \boldsymbol{y}] = -[\boldsymbol{y}, \boldsymbol{x}]. \tag{8.1.8}$$

4. The Jacobi identity holds

$$[\boldsymbol{x}, [\boldsymbol{y}, \boldsymbol{z}]] + [\boldsymbol{y}, [\boldsymbol{z}, \boldsymbol{x}]] + [\boldsymbol{z}, [\boldsymbol{x}, \boldsymbol{y}]] = 0. \tag{8.1.9}$$

If the field K is the set of real numbers we say that $\mathfrak{g}$ is a REAL LIE ALGEBRA. *If K is the field of complex numbers $\mathfrak{g}$ is then called a* COMPLEX LIE ALGEBRA. ■

If $\{\boldsymbol{e}_i\}$ is a basis of a Lie algebra then by the property 1 there must hold

$$[\boldsymbol{e}_i, \boldsymbol{e}_j] = \sum_k c_{ijk} \boldsymbol{e}_k. \tag{8.1.10}$$

The constants c_{ijk} are called the STRUCTURE CONSTANTS *of the Lie algebra.* Consequently a Lie algebra can be characterized by a set of constants c_{ijk} such that by (8.1.8) and (8.1.9)

$$\begin{aligned} &c_{ijk} = -c_{jik}, \\ &\textstyle\sum_m c_{ijm}c_{mkr} + c_{jkm}c_{mir} + c_{kim}c_{mjr} = 0, \end{aligned} \qquad r = 1, \ldots, n. \tag{8.1.11}$$

The Lie Algebra Associated with a Lie Group

We shall admit that with every Lie group $\boldsymbol{G}$ we can associate in an intrinsic way a Lie algebra $\mathfrak{g}$ called the Lie algebra of the group.

The associated Lie algebra is formed by the infinitesimal generators of the Lie group or by its infinitesimal matrices in the case where the dimension is finite. In particular, we have seen that the infinitesimal generators of a Lie group are related to each other by relations (2.1.34) where the structure constants appear, these relations being the same kind as (8.1.10).

Small gothic letters will be used to denote the Lie algebra associated with a group, the latter being denoted by the capital italic letters of those used for its algebra. For example, the Lie algebra $\mathfrak{su}(2)$ is associated with the Lie group $\boldsymbol{SU}(2)$.

8.1.3 Isomorphic Lie Algebras

The vector space $\mathbb{R}^3$ provided with the vector product as a Lie bracket, $[\boldsymbol{x}, \boldsymbol{y}] = (\boldsymbol{x} \times \boldsymbol{y})$, is an example of a Lie algebra. If $\{\boldsymbol{e}_i\}$ is an orthonormal basis of $\mathbb{R}^3$ in the usual sense we have

$$\begin{aligned} \boldsymbol{e}_j \times \boldsymbol{e}_k &= [\boldsymbol{e}_j, \boldsymbol{e}_k] \\ &= \sum_m \varepsilon_{jkm} \boldsymbol{e}_m, \end{aligned} \tag{8.1.12}$$

where ε_{jkm} is the antisymmetric symbol. The quantities ε_{jkm} are thus the structure constants of this Lie algebra.

We have seen that the infinitesimal matrices L_i of the groups $\boldsymbol{SO}(3)$ also form a basis of a Lie algebra $\mathfrak{so}(3)$, and that they are related by the relations

$$[L_j, L_k] = \sum_m \varepsilon_{jkm} L_m. \tag{8.1.13}$$

Equations (8.1.12) and (8.1.13) show that the two algebras $\mathbb{R}^3$ and $\mathfrak{so}(3)$ are defined by the same structure constants. They are isomorphic Lie algebras.

Let us also consider the following real Lie algebra having as its basis the matrices

$$M_1 = \frac{1}{2}\begin{bmatrix} 0 & -i \\ -i & 0 \end{bmatrix}, \qquad M_2 = \frac{1}{2}\begin{bmatrix} 0 & -1 \\ -1 & 0 \end{bmatrix},$$
$$M_3 = \frac{1}{2}\begin{bmatrix} -i & 0 \\ 0 & i \end{bmatrix}. \tag{8.1.14}$$

These matrices are given by (1.3.7), (1.3.9), and (1.3.11). They are the infinitesimal matrices of the groups $\boldsymbol{SU}(2)$, and the Lie algebra with these matrices as its basis is denoted $\mathfrak{su}(2)$. This algebra is also isomorphic to the algebras $\mathbb{R}^3$ and $\mathfrak{so}(3)$.

Let us remark that a Lie algebra is called real if it is defined over the field of real numbers. That does not mean that the matrices of this algebra do not contain complex numbers, as is shown by the matrices (8.1.14).

Isomorphic Lie Algebras

Two Lie algebras $\mathfrak{g}_1$ and $\mathfrak{g}_2$ will be isomorphic if there exists an invertible mapping f of $\mathfrak{g}_1$ into $\mathfrak{g}_2$ such that

$$f(\alpha\boldsymbol{x} + \beta\boldsymbol{y}) = \alpha f(\boldsymbol{x}) + \beta f(\boldsymbol{y}), \tag{8.1.15}$$

$$[f(\boldsymbol{x}), f(\boldsymbol{y})]_2 = f([\boldsymbol{x}, \boldsymbol{y}]_1), \tag{8.1.16}$$

where $[\ ,\ ]_1$ and $[\ ,\ ]_2$ denote the Lie brackets in $\mathfrak{g}_1$ and $\mathfrak{g}_2$, respectively. We easily verify that this definition of isomorphic Lie algebras applies to the examples given above.

8.2 Representations of Lie Algebras

8.2.1 Definition

Let us consider a Lie algebra $\mathcal{L}$ and let us assume that with each element $\boldsymbol{a}$ of $\mathcal{L}$ we can associate a square matrix of order n, denoted $M(\boldsymbol{a})$, which is such that

$$M(\alpha \boldsymbol{a} + \beta \boldsymbol{b}) = \alpha M(\boldsymbol{a}) + \beta M(\boldsymbol{b}), \tag{8.2.1}$$

where α and β are real or complex numbers according as $\mathcal{L}$ is a real or complex Lie algebra. Furthermore

$$M([\boldsymbol{a}, \boldsymbol{b}]) = [M(\boldsymbol{a}), M(\boldsymbol{b})]. \tag{8.2.2}$$

We then say that these matrices form an n-dimensional representation Γ_n of the Lie algebra $\mathcal{L}$. In fact, the set of matrices $M(\boldsymbol{a})$ of this representation is a Lie algebra $\mathcal{R}$ defined over the same field as $\mathcal{L}$. In order to define a representation it is sufficient to define the matrices $M(\boldsymbol{e}_i)$ of each element $\boldsymbol{e}_i$ of the basis of $\mathcal{L}$.

The classical notions of representations of finite groups are recovered for the representations of Lie algebras. Thus, if S is a regular matrix of order n and if Γ_n is an n-dimensional representation of $\mathcal{L}$, then the set of matrices $M'(\boldsymbol{a})$ defined by

$$M'(\boldsymbol{a}) = S^{-1} M(\boldsymbol{a}) S \tag{8.2.3}$$

is also a representation $\Gamma_n{}'$ of $\mathcal{L}$. The representations Γ_n and $\Gamma_n{}'$ are equivalent representations.

The notions of reducible or irreducible representation are also carried over to Lie algebras. Thus, a representation Γ of $\mathcal{L}$ will be called completely reducible if it is equivalent to a representation Γ' of $\mathcal{L}$ for which the matrices of Γ' have the form

$$M'(\boldsymbol{a}) = \begin{bmatrix} M_1{}'(\boldsymbol{a}) & & \dots & 0 \\ 0 & M_2{}'(\boldsymbol{a}) & \dots & 0 \\ \vdots & \vdots & \vdots & \vdots \\ 0 & 0 & \dots & M_k{}'(\boldsymbol{a}) \end{bmatrix} \tag{8.2.4}$$

for every element $\boldsymbol{a}$ of $\mathcal{L}$ and where the matrices $M_i{}'(\boldsymbol{a})$ define irreducible representations $\Gamma_i{}'$ of $\mathcal{L}$.

Examples of Representations

The Lie algebra $\mathfrak{so}(2)$ is the one which corresponds to the Lie group $\boldsymbol{SO}(2)$ of plane rotations. It is a one-dimensional algebra and we can take as the basis element the matrix

$$\boldsymbol{a}_1 = \begin{bmatrix} 0 & -1 \\ -1 & 0 \end{bmatrix}. \tag{8.2.5}$$

The only commutation relation of this algebra is $[\boldsymbol{a}_1, \boldsymbol{a}_1] = 0$. A one-dimensional representation of $\mathfrak{so}(2)$ is given by

$$M_{\mathcal{L}}(\boldsymbol{a}_1) = p,$$

where p is an arbitrary complex number.

As another example let us take that of the Lie algebra $\mathfrak{so}(3)$ which, as we have seen, is isomorphic to the real Lie algebra $\mathfrak{su}(2)$. Thus a two-dimensional representation of $\mathfrak{so}(3)$ is given by the matrices of $\mathfrak{su}(2)$, namely,

$$M_{\mathcal{L}}(\boldsymbol{a}_1) = \frac{1}{2}\begin{bmatrix} 0 & -i \\ -i & 0 \end{bmatrix}, \qquad M_{\mathcal{L}}(\boldsymbol{a}_2) = \frac{1}{2}\begin{bmatrix} 0 & -1 \\ 1 & 0 \end{bmatrix},$$
$$M_{\mathcal{L}}(\boldsymbol{a}_3) = \frac{1}{2}\begin{bmatrix} -i & 0 \\ 0 & i \end{bmatrix}, \tag{8.2.6}$$

where $\boldsymbol{a}_1, \boldsymbol{a}_2, \boldsymbol{a}_3$ are elements of the basis of $\mathfrak{so}(3)$.

8.2.2 Representations of a Lie Group and of Its Lie Algebra

The relation between the linear representation of a Lie group and that of its associated Lie algebra is particularly important. The procedure given above always allows us to go from the representation of the group to that of its Lie algebra. In contrast, it does not always hold in the reverse direction.

THEOREM **8.1.** *Let $\Gamma_{\boldsymbol{G}}$ be an n-dimensional representation of a Lie group given by the matrices $M_{\boldsymbol{G}}$, and the associated real Lie algebra of which is denoted $\mathcal{L}$. Then there exists an n-dimensional representation $\Gamma_{\mathcal{L}}$ of the Lie algebra $\mathcal{L}$ defined for every element $\boldsymbol{a}$ of $\mathcal{L}$ by the following matrices*

$$M_{\mathcal{L}}(\boldsymbol{a}) = \frac{\mathrm{d}}{\mathrm{d}t}\,[M_{\boldsymbol{G}}[\exp(t\boldsymbol{a})]_{t=0}. \tag{8.2.7}$$

For every $\boldsymbol{a}$ belonging to $\mathcal{L}$ and for every real t we have

$$\exp[tM_{\mathcal{L}}(\boldsymbol{a})] = M_{\boldsymbol{G}}[\exp(t\boldsymbol{a})]. \tag{8.2.8}$$

■

We shall admit this theorem and give two illustrations of it showing its limits of applicability.

EXAMPLE **8.2.** Let us return to the examples of the Lie algebra $\mathfrak{so}(2)$ having as its basis element the matrix $\boldsymbol{a}_1$ given by (8.2.5). According to (8.2.8) we obtain as a representation of $\mathfrak{so}(2)$

$$\exp[tM_{\mathcal{L}}(\boldsymbol{a}_1)] = \exp(tp). \tag{8.2.9}$$

On the other hand, the representation matrix of the group $\boldsymbol{SO}(2)$ of plane rotations has as its expression

$$\exp(t\boldsymbol{a}_1) = \begin{bmatrix} \cos t & -\sin t \\ \sin t & \cos t \end{bmatrix}. \tag{8.2.10}$$

It is noted that for (8.2.10) we have

$$\exp[(t+2\pi)\boldsymbol{a}_1] = \exp(t\boldsymbol{a}_1),$$

then the representation (8.2.9) gives us

$$\exp[(t+2\pi)M_{\mathcal{L}}(\boldsymbol{a}_1)] = \exp(2\pi p)[tM_{\mathcal{L}}(\boldsymbol{a}1)]. \tag{8.2.11}$$

Consequently this matrix representation $M_{\mathcal{L}}(\boldsymbol{a}_1)$ of the Lie algebra $\mathfrak{so}(2)$ gives a representation of the group $\boldsymbol{SO}(2)$ by exponentiation if and only if $p = iq$, where q is an integer. ■

EXAMPLE **8.3.** Let $\boldsymbol{a}_1, \boldsymbol{a}_2, \boldsymbol{a}_3$ be the matrices forming a basis of the Lie algebra $\mathfrak{so}(3)$. A two-dimensional representation of this algebra is given by the matrices $M_{\mathcal{L}}(\boldsymbol{a}_i)$, $i = 1, 2, 3$. By exponentiation we obtain a matrix representation of $\boldsymbol{SO}(3)$, thus, for example, for $M_{\mathcal{L}}(\boldsymbol{a}_3)$

$$\exp[tM_{\mathcal{L}}(\boldsymbol{a}_3)] = \begin{bmatrix} e^{it/2} & 0 \\ 0 & e^{-it/2} \end{bmatrix}. \tag{8.2.12}$$

On the other hand, for the matrix corresponding to a rotation about the Oz axis we have

$$\exp(t\boldsymbol{a}_3) = \begin{bmatrix} \cos t & -\sin t & 0 \\ \sin t & \cos t & 0 \\ 0 & 0 & 1 \end{bmatrix}. \tag{8.2.13}$$

We notice that we have, on the one hand

$$\exp[(t+2\pi)\boldsymbol{a}_3] = -\exp(t\boldsymbol{a}_3),$$

and, on the other,

$$\exp[(t+2\pi)M_{\mathcal{L}}(\boldsymbol{a}_3)] = -\exp[tM_{\mathcal{L}}(\boldsymbol{a}_3)]. \tag{8.2.14}$$

Consequently, the matrix representation $M_{\mathcal{L}}(\boldsymbol{a}_3)$ of $\mathfrak{so}(3)$ does not give us a representation of $\boldsymbol{SO}(3)$ by exponentiation. ■

8.2.3 Connected Groups

Let us give some idea of the notion of connected group by considering the case of the group $\boldsymbol{SO}(2)$. It is a one-parameter group and it is possible to identify the elements of this group with the points of a circle of radius ρ by associating a rotation through an angle θ with a point M of the curvilinear abscissa $\rho\theta$ measured from a point of origin. We note that this correspondence is bijective and respects the topology of the group in the sense that two neighboring rotations are associated with two neighboring points of the circle. We say that the circle is a CONNECTED SPACE because we can move from one point of the circle to another in a continuous manner. It will be said that the group $\boldsymbol{SO}(2)$ has the topology of the circle, and $\boldsymbol{SO}(2)$ is called a CONNECTED GROUP.

The same can be done for plenty of groups in making them correspond to a connected space, and these groups will be called connected groups. Thus we can associate with each matrix of $\boldsymbol{SU}(2)$ a point on a sphere in four-dimensional space. In fact, the matrices of $\boldsymbol{SU}(2)$ have two complex parameters

$$a = x + iy \qquad \text{and} \qquad b = z + it$$

related by

$$|a|^2 + |b|^2 = 1. \tag{8.2.15}$$

We can then associate with a matrix of $\boldsymbol{SU}(2)$ a point on a sphere with equation

$$x^2 + y^2 + z^2 + t^2 = 1. \tag{8.2.16}$$

As the surface of a sphere is a connected space the group $\boldsymbol{SU}(2)$ is called a connected group. Furthermore, we prove that all the groups $\boldsymbol{SU}(n)$ are connected and that the groups $\boldsymbol{SO}(n)$ are also.

In contrast, the group $\boldsymbol{O}(2)$, for example, is not a connected group because we cannot 'move' in a continuous manner from one rotation to a symmetry with respect to a line. The topological space with which $\boldsymbol{O}(2)$ can be associated is made up to two distinct circles, one corresponding to the subgroup $\boldsymbol{SO}(2)$ and the other to inversion rotations.

8.2.4 Reducible and Irreducible Representations

THEOREM **8.2.** *Let $\Gamma_{\boldsymbol{G}}$ be an n-dimensional representation of a Lie group $\boldsymbol{G}$ of which the associated Lie algebra is $\mathcal{L}$. An n-dimensional representation of $\mathcal{L}$ is denoted $\Gamma_{\mathcal{L}}$.*

— *If $\Gamma_{\boldsymbol{G}}$ is a reducible representation then $\Gamma_{\mathcal{L}}$ is reducible. The converse is also true if $\boldsymbol{G}$ is a connected group.*

— *The representation $\Gamma_{\mathcal{L}}$ is completely reducible if $\Gamma_{\boldsymbol{G}}$ is. The converse is true if $\boldsymbol{G}$ is connected.*

— *If $\boldsymbol{G}$ is a connected group then $\Gamma_{\mathcal{L}}$ is irreducible if $\Gamma_{\boldsymbol{G}}$ is irreducible. Conversely, $\Gamma_{\boldsymbol{G}}$ is irreducible if $\Gamma_{\mathcal{L}}$ also is irreducible.* ■

Proof. The proof of this theorem follows from (8.2.7). In fact, if $\Gamma_{\boldsymbol{G}}$ is reducible, by differentiation of the matrix elements we will then obtain matrices with the same structure as the representation $\Gamma_{\mathcal{L}}$. The same is true when $\Gamma_{\boldsymbol{G}}$ is completely reducible.

Conversely, let us consider the case of a diagonal matrix of blocks (the matrix (8.2.4)) corresponding to a completely reducible representation. The different powers of a quasi-diagonal matrix have the same structure, and the exponential matrix formed by the sum of theses different matrices will also be a quasi-diagonal matrix. The argument is the same for a reducible matrix representation.

Lastly, the relation between the irreducibility of the representations $\Gamma_{\boldsymbol{G}}$ and $\Gamma_{\mathcal{L}}$ results from the preceding properties. ■

8.3 Clifford Algebras

8.3.1 Definition

Let us consider n elements $\boldsymbol{e}_j$, $1 \leqslant j \leqslant n$, satisfying the conditions

$$\begin{aligned} \boldsymbol{e}_i \boldsymbol{e}_j + \boldsymbol{e}_j \boldsymbol{e}_i &= \{\boldsymbol{e}_i, \boldsymbol{e}_j\} \\ &= 2\delta_{ij}. \end{aligned} \tag{8.3.1}$$

The elements $\boldsymbol{e}_j$ can be, for example, matrices or operators. From the generators of these n elements $\boldsymbol{e}_j$, called GENERATING ELEMENTS, we form the following products

$$\boldsymbol{e}_{i_1} \boldsymbol{e}_{i_2} \ldots \boldsymbol{e}_{i_p}, \tag{8.3.2a}$$

with $i_1 < i_2 < \cdots < i_p$ and $0 \leqslant p \leqslant n$. When p varies from 0 to n the number of combinations obtained is thus equal to

$$\sum_{p=0}^{n} C_p^n, \tag{8.3.2b}$$

and by Newton's binomial formula this sum is equal to $(1+1)^n$. We thus obtain 2^n different elements on adding the unit element to these products, that is to say,

$$1, \boldsymbol{e}_i,\ \boldsymbol{e}_i \boldsymbol{e}_j,\ \ldots,\ \boldsymbol{e}_1 \boldsymbol{e}_2 \ldots \boldsymbol{e}_n. \tag{8.3.3}$$

Since the $\boldsymbol{e}_j$ anticommute, $\boldsymbol{e}_i \boldsymbol{e}_j = -\boldsymbol{e}_j \boldsymbol{e}_i$, we can always arrange their products so that the factors appear in increasing order.

EXAMPLE **8.4.** By way of an example let us consider, for $n = 2$, the matrices

$$e_1 = \begin{bmatrix} 0 & 1 \\ 1 & 0 \end{bmatrix}, \qquad e_3 = \begin{bmatrix} 0 & i \\ -i & 0 \end{bmatrix}. \tag{8.3.4}$$

We verify that

$$(e_1)^2 = (e_2)^2 = 1 \qquad \text{and} \qquad \{e_1, e_2\} = 0.$$

The elements $1, e_1, e_2, e_1e_2$ are linearly independent and are able to serve as a basis of a vector space of dimension $2^n = 4$. ■

DEFINITION **8.2.** *Quite generally, it is proved that the* 2^n *products* (8.3.3) *are linearly independent and thus form a basis of a vector space over the field of reals. The notion of product provides this vector space with the structure of an algebra which is called a* CLIFFORD ALGEBRA, *and which is denoted* $\boldsymbol{C}_n$. ■

Let us remark that instead of taking (8.3.1) as the condition, we can also choose the following properties

$$e_ie_j = -e_je_i \qquad \text{if } i \neq j \quad \text{and} \quad e_j^2 = -1.$$

To form the relation between the two definitions it suffices to multiply the elements e_j by the complex number i.

Even and Odd Subalgebras

Let us consider the products $e_{i_1}e_{i_2}\dots e_{i_p}$ of a Clifford algebra $\boldsymbol{C}_n$ and let us choose only the products where p is *even*. These products, with the unit element, are 2^{n-1} in number. They form a basis of a (2^{n-1})-dimensional EVEN SUBALGEBRA, denoted C_{+n}, of the algebra $\boldsymbol{C}_n$. The products $e_{i_1}e_{i_2}\dots e_{i_p}$, with p odd, form an ODD SUBALGEBRA denoted C_{-n}.

8.3.2 Examples of Clifford Algebras

Pauli Matrices

The Pauli matrices σ_i satisfy the relation

$$\{\sigma_i, \sigma_j\} = 2\delta_{ij} \qquad \text{for} \quad i, j = 1, 2, 3. \tag{8.3.5}$$

These matrices can therefore serve as generating elements of a Clifford algebra. Let us write σ_0 for the unit matrix of order 2. We can then form the following $2^3 = 8$ matrices

$$\sigma_0,\ \sigma_1,\ \sigma_2,\ \sigma_3,\ \sigma_1\sigma_2,\ \sigma_2\sigma_3,\ \sigma_2\sigma_3,\ \sigma_1\sigma_2\sigma_3. \tag{8.3.6}$$

Furthermore, the Pauli matrices satisfy the following relations

$$\sigma_1\sigma_2 = i\sigma_2, \qquad \sigma_2\sigma_3 = i\sigma_1, \qquad \sigma_3\sigma_1 = i\sigma_2, \qquad \sigma_1\sigma_2\sigma_3 = i\sigma_0, \quad (8.3.7)$$

and we obtain the eight following matrices

$$\sigma_0,\ \sigma_1,\ \sigma_2,\ \sigma_3,\ i\sigma_3,\ i\sigma_1,\ i\sigma_2,\ i\sigma_0,$$

which are linearly independent over the field of reals. As the real algebra, denoted $\boldsymbol{C}(E_3)$, has the Pauli matrices as its generating elements it is therefore an eight-dimensional Clifford algebra.

As the three Pauli matrices are linearly independent they are able to form a basis of a three-dimensional vector space V_3. In Chapter 1 we have seen that to every vector $\boldsymbol{x} = (x, y, z)$ of the Euclidean space E_3 we can make correspond a Hermitian matrix X by (1.3.29) such that

$$X = x\sigma_1 + y\sigma_2 + z\sigma_3.$$

This correspondence is an isomorphism between the usual Euclidean vector space E_3 and the vector space E_3 generated by the Pauli matrices. We shall then say that the real algebra $\boldsymbol{C}(E_3)$ is a Clifford algebra of the Euclidean space E_3.

As the algebra has the four matrices $\sigma_0, i\sigma_3, i\sigma_1, i\sigma_2$ as basis elements it forms a subalgebra, denoted $\boldsymbol{C}_+(E_3)$, of the Clifford algebra $\boldsymbol{C}(E_3)$.

Quaternions

The mathematician Hamilton thought up *quaternions* in the last century in order to generalize the notion of complex number. To do so Hamilton introduced three symbols $\boldsymbol{i}, \boldsymbol{j}, \boldsymbol{k}$ with the following properties

$$\begin{aligned} \boldsymbol{i}^1 = \boldsymbol{j}^2 = \boldsymbol{k}^2 = -1, &\qquad \boldsymbol{ij} = -\boldsymbol{ji} = -\boldsymbol{k}, \\ \boldsymbol{jk} = -\boldsymbol{kj} = -\boldsymbol{i}, &\qquad \boldsymbol{ki} = -\boldsymbol{ik} = -\boldsymbol{j}. \end{aligned} \quad (8.3.8)$$

From these entities and unity, 1, we can form arbitrary combinations over the field of reals having the general form

$$h = a1 + b\boldsymbol{i} + c\boldsymbol{j} + d\boldsymbol{k}. \quad (8.3.9)$$

The entities h formed thus are called quaternions and they are able to form a four-dimensional vector space over the field of reals. We can then provide this vector space with the structure of an algebra. We then have obtained the QUATERNION ALGEBRA, denoted $\boldsymbol{H}$.

The quaternion algebra is a Clifford algebra, ie by taking

$$\boldsymbol{e}_i\boldsymbol{e}_j = -\boldsymbol{e}_j\boldsymbol{e}_i \qquad \text{for} \quad i \neq j \quad \text{and} \quad (\boldsymbol{e}_j)^2 = -1.$$

The generating elements are $\boldsymbol{e}_1 = \boldsymbol{i}$ and $\boldsymbol{e}_2 = \boldsymbol{j}$. Their product gives the third element

$$\boldsymbol{e}_3 = \boldsymbol{e}_1\boldsymbol{e}_2 = \boldsymbol{ij} = -\boldsymbol{k}.$$

On the other hand, let us consider the Pauli matrices, each multiplied by i. Adding σ_0 to the matrices $i\sigma_j$ we obtain four matrices which have the same multiplication table as the symbols $\boldsymbol{i}, \boldsymbol{j}, \boldsymbol{k}$ and unity. A matrix representation of the quaternions is then the following

$$h = a\sigma_0 + b(i\sigma_1) + c(i\sigma_2) + d(i\sigma_3). \tag{8.3.10}$$

The quaternion algebra $\boldsymbol{H}$ can thus be identified with the subalgebra $\boldsymbol{C}_+(E_3)$ generated by the matrix of (8.3.10). If we write α_j, $j = 1, 2, 3$, for the generating symbols of the quaternions, we note that the elements $\beta_j i\alpha_j$ satisfy the defining property

$$\{\beta_j, \beta_i\} = 2\delta_{ij}$$

of the generating elements of a Clifford algebra. This property allows us to find an isomorphism between the quaternion algebra and other four-dimensional algebras.

A classical representation of the quaternions is that obtained from the following generator elements and their product

$$\boldsymbol{e}_1 = \begin{bmatrix} 0 & 1 \\ -1 & 0 \end{bmatrix}, \qquad \boldsymbol{e}_2 = \begin{bmatrix} 0 & i \\ i & 0 \end{bmatrix},$$
$$\boldsymbol{e}_3 = \boldsymbol{e}_1\boldsymbol{e}_2 = \begin{bmatrix} i & 0 \\ 0 & -1 \end{bmatrix}. \tag{8.3.11}$$

We then obtain for the matrix representation of a quaternion

$$h = a\mathbb{1} + b\boldsymbol{e}_1 + c\boldsymbol{e}_2 + d\boldsymbol{e}_3. \tag{8.3.12}$$

Dirac Matrices

The Dirac matrices are super-matrices formed from Pauli matrices. There are four Dirac matrices

$$\gamma_i = \begin{bmatrix} 0 & \sigma_j \\ -\sigma_j & 0 \end{bmatrix}, \quad j = 1, 2, 3, \qquad \gamma_0 = \begin{bmatrix} \mathbb{1} & 0 \\ 0 & -\mathbb{1} \end{bmatrix}. \tag{8.3.13}$$

The Dirac matrices satisfy the following relations

$$\{\gamma_j, \gamma_k\} = 2g_{jk}\mathbb{1}, \tag{8.3.14}$$

where g_{jk} is the metric tensor of Minkowski space, chosen in the present case to be such that

$$g_{jk} = 0 \quad \text{if } j \neq k, \qquad g_{00} = 1, \qquad g_{kk} = -1. \tag{8.3.15}$$

Thus we obtain for the Dirac matrices,

$$(\gamma_0)^2 = \mathbb{1}, \qquad (\gamma_k)^2 = -\mathbb{1}.$$

Replacing γ_0 by α_0 and γ_k by $i\alpha_k$ equations (8.3.14) become

$$\{\alpha_j, \alpha_k\} = 2\delta_{jk}\mathbb{1}, \qquad i, j = 1, 2, 3, 4. \tag{8.3.16}$$

The matrices α_k allow us to obtain $2^4 = 16$ product elements which are linearly independent and form a basis of a sixteen-dimensional Clifford algebra, denoted $\boldsymbol{C}(E_{1,3})$.

We can identify Minkowski space, denoted $E_{1,3}$, with the vector space of Hermitian matrices generated by $\sigma_0, \sigma_1, \sigma_2, \sigma_3$. To a vector $\boldsymbol{X} = (x_0, x_1, x_2, x_3)$ of Minkowski space there corresponds the Hermitian matrix X defined by

$$X = x_0\sigma_0 + x_1\sigma_1 + x_2\sigma_2 + x_3\sigma_3. \tag{8.3.17}$$

We shall say that the sixteen-dimensional real algebra $\boldsymbol{C}(E_{1,3})$ is the Clifford algebra of the Minkowski space $E_{1,3}$.

The Brauer–Weyl Construction

This construction generalizes those of the algebras $\boldsymbol{C}(E_3)$ and $\boldsymbol{C}(E_{1,3})$ which have been obtained from the Pauli matrices.

— For even n, $n = 2p$, we obtain generating elements of a Clifford algebra on taking the following elements

$$\sigma_j = \overbrace{\sigma_3 \otimes \cdots \otimes \sigma_3}^{j-1} \otimes \sigma_1 \otimes \overbrace{\mathbb{1} \otimes \cdots \otimes \mathbb{1}}^{p-1}, \qquad j \leqslant p, \tag{8.3.18}$$

where the matrix σ_3 is tensorially multiplied with itself $j-1$ times, the matrix σ_1 is located at the jth place, and the matrix $\mathbb{1}$ is tensorially multiplied with itself $p-1$ times. On the other hand, the elements σ_{j+p} have as their expressions

$$\alpha_{j+p} = \overbrace{\sigma_3 \otimes \cdots \otimes \sigma_3}^{j-1} \otimes \sigma_2 \otimes \overbrace{\mathbb{1} \otimes \cdots \otimes \mathbb{1}}^{p-1}, \qquad j \leqslant p. \tag{8.3.19}$$

— For odd n, $n = 2p + 1$, the matrix σ_{2p+1} is given by

$$\alpha_{2p+1} = \overbrace{\sigma_3 \otimes \cdots \otimes \sigma_3}^{p}, \tag{8.3.20}$$

where the matrix σ_3 is tensorially multiplied with itself p times.

For $n = 3$ we recover the generating elements of the Clifford algebra $\boldsymbol{C}(E_3)$ with $\alpha_1 = \sigma_1$, $\sigma_1 = \sigma_3$, $\alpha_3 = \sigma_3$.

For $n = 4$ we have the following generating elements

$$\begin{aligned} \alpha_1 = \sigma_1 \otimes \mathbb{1} = \begin{bmatrix} \sigma_1 & 0 \\ 0 & \sigma_1 \end{bmatrix}, \quad \alpha_2 = \sigma_3 \otimes \sigma_1 = \begin{bmatrix} 0 & \sigma_3 \\ \sigma_3 & 0 \end{bmatrix}, \\ \alpha_3 = \sigma_2 \otimes \mathbb{1} = \begin{bmatrix} \sigma_2 & 0 \\ 0 & \sigma_2 \end{bmatrix}, \quad \alpha_4 = \sigma_3 \otimes \sigma_1 = \begin{bmatrix} 0 & -i\sigma_3 \\ i\sigma_3 & 0 \end{bmatrix}, \end{aligned} \tag{8.3.21}$$

It is easily verified that the matrices satisfy the anticommutation relation

$$\{\alpha_i, \sigma_j\} = 2\delta_{ij}.$$

For example

$$\begin{aligned} \{\alpha_1, \alpha_2\} &= \sigma_1\sigma_3 \otimes \sigma_1 + \sigma_3\sigma_1 \otimes \sigma_1 \\ &= \{\sigma_1, \sigma_3\} \otimes \sigma_1 \\ &= 0. \end{aligned} \tag{8.3.22}$$

The matrices (8.3.21) are not the classical Dirac matrices but can be obtained by multiplying together the latter. We obtain

$$\alpha_1 = i\gamma_2\gamma_3, \qquad \alpha_2 = -\gamma_3\gamma_0, \qquad \alpha_3 = i\gamma_3\gamma_1, \qquad \alpha_4 = -i\gamma_3. \tag{8.3.23}$$

Since the basis elements of a Clifford algebra are obtained by multiplying together some of its generating elements, the elements α_i given by (8.3.21) also allow us to obtain the Clifford algebra $C(E_{1,3})$ of Minkowski space.

8.3.3 Clifford and Lie Algebras

The Lie Algebra $\mathfrak{so}(3)$

Let us consider three elements $\alpha_1, \alpha_2, \alpha_3$ forming the generators of a Clifford algebra and which thus satisfy the anticommutation relations (8.3.1). We are going to show that the elements

$$M_{ij} = \tfrac{1}{2}\alpha_i\alpha_j$$

form a basis of the Lie algebra $\mathfrak{so}(3)$. To do so we are going to calculate the commutators of M_{ij} and show that we obtain exactly the commutation relations of the elements forming a basis of $\mathfrak{so}(3)$.

Consequently by the anticommutation property we have only three independent elements M_{ij}, that is to say,

$$\begin{aligned} M_{12} &= \tfrac{1}{2}\alpha_1\alpha_2 = -M_{21}, \\ M_{13} &= \tfrac{1}{2}\alpha_1\alpha_3 = -M_{31}, \\ M_{23} &= \tfrac{1}{2}\alpha_2\alpha_3 = -M_{32}. \end{aligned} \tag{8.3.24}$$

Taking into account the anticommutation property of the element α_j, the commutator $[M_{12}, M_{13}]$ may be written

$$\begin{aligned}
[M_{12}, M_{13}] &= \tfrac{1}{4}(\alpha_1\alpha_2\alpha_1\alpha_3 - \alpha_1\alpha_3\alpha_1\alpha_2) \\
&= M_{32},
\end{aligned} \tag{8.3.25}$$

where we have used the relation

$$\begin{aligned}
\alpha_1\alpha_2\alpha_1\alpha_3 &= \alpha_1(2\delta_{21} - \alpha_1\alpha_2)\alpha_3 \\
&= -\alpha_1\alpha_1\alpha_2\alpha_3 \\
&= 2M_{32}.
\end{aligned} \tag{8.3.26}$$

The two other commutators are calculated similarly, and we obtain

$$[M_{32}, M_{12}] = M_{13}, \qquad [M_{13}, M_{32}] = M_{12}. \tag{8.3.27}$$

If we write

$$M_{12} = L_x, \qquad M_{13} = L_y, \qquad M_{32} = L_z,$$

it is seen that the commutation relations (8.3.25) and (8.3.27) are exactly those which characterize the basis elements of the Lie algebra $\mathfrak{so}(3)$. These elements are the infinitesimal generators of the group of spacial rotations $\boldsymbol{SO}(3)$. These commutation relations determine the structure constants and thus define, to within an isomorphism, a Lie algebra.

The Lie Algebra $\mathfrak{so}(n)$

The preceding result is easily generalized to the Lie algebra $\mathfrak{so}(n)$ associated with the group $\boldsymbol{SO}(n)$ of rotations in an n-dimensional space.

To do so let us consider a set of elements α_j, $j = 1, \ldots, n$, satisfying the anticommutation relations (8.3.1). The elements $M_{ij} = \frac{1}{2}\alpha_i\alpha_j$ then form a basis of the Lie algebra $\mathfrak{so}(n)$. As before, the calculation of the commutators

$$[M_{ij}, M_{rs}] = \tfrac{1}{4}(\alpha_i\alpha_j\alpha_r\alpha_s - \alpha_r\alpha_s\alpha_i\alpha_j) \tag{8.3.28}$$

uses the anticommutation property of the elements α_j, as was done in (8.3.26), and we obtain

$$[M_{ij}, M_{rs}] = \delta_{is}M_{jr} + \delta_{ir}M_{sj} + \delta_{sj}M_{ri} + \delta_{rj}M_{is}. \tag{8.3.29}$$

These are precisely the commutation relations of the matrices M_{jk} which generate the orthogonal group $\boldsymbol{O}(n)$.

Let us look for a representation of this group by considering the vector space V_n generated by the elements α_k as the representation space. The matrices

$$S_{ij}(\theta) = \exp(-\theta M_{ij}),$$

when taking into account the anticommutation relations of the α_k, have the expressions

$$\begin{aligned} S_{ij}(\theta) &= \exp[-(\tfrac{1}{2}\theta)\alpha_i\alpha_j] \\ &= \cos(\tfrac{1}{2}\theta)\mathbb{1} - \alpha_i\alpha_j \sin(\tfrac{1}{2}\theta). \end{aligned} \tag{8.3.30}$$

Let us consider an arbitrary vector $\boldsymbol{x}$ of V_n and let us show that the operation $S_{ij}(\theta)\boldsymbol{x}S_{ij}(-\theta)$ is a rotation. For example, the operation $S_{ij}(\theta)\alpha_i \times S_{ij}(-\theta)$ gives us

$$\begin{aligned} S_{ij}(\theta)\alpha_i S_{ij}(-\theta) &= \alpha_i \cos\theta + \alpha_j \sin\theta, \\ S_{ij}(\theta)\alpha_j S_{ij}(-\theta) &= -\alpha_i \sin\theta + \alpha_j \cos\theta, \end{aligned} \tag{8.3.31}$$

and all the other vectors stay fixed. We certainly obtain a rotation in V_n, and the matrices

$$S_{ij}(\theta) = \exp(-\theta M_{ij})$$

form a representation of $\boldsymbol{SO}(n)$. The matrices M_{ij} thus constitute a basis of the Lie algebra $\mathfrak{so}(n)$.

8.3.4 Spinor Groups

Three-Dimensional Euclidean Space

To every vector

$$\boldsymbol{x} = (x_1, x_2, x_3) \qquad \text{and} \qquad \boldsymbol{y} = (y_1, y_2, y_3)$$

of the usual Euclidean space E_3 we can bring into correspondence the following matrices

$$X = x_1\sigma_1 + x_2\sigma_2 + x_3\sigma_3, \qquad Y = y_1\sigma_1 + y_2\sigma_2 + y_3\sigma_3, \tag{8.3.32}$$

where the σ_j are the Pauli matrices. The product of these matrices may be written

$$XY = (\boldsymbol{x}\cdot\boldsymbol{y})\sigma_0 + i\{(x_2y_3 - x_3y_2)\sigma_1 + (x_3y_1 - x_1y_3)\sigma_2 + (x_1y_2 - x_2y_1)\sigma_3\}, \tag{8.3.33}$$

where σ_0 is the unit matrix. Let us write the vector product of $\boldsymbol{x}$ with $\boldsymbol{y}$ as

$$\boldsymbol{z} = \boldsymbol{x} \times \boldsymbol{y} = (z_1, z_2, z_3)$$

and let Z be the matrix corresponding to $\boldsymbol{z}$, and let

$$\rho = \boldsymbol{x}\cdot\boldsymbol{y}.$$

The matrix product (8.3.33) is then written

$$XY = \rho\sigma_0 + z_1 i\sigma_1 + z_2 i\sigma_2 + x_3 i\sigma_3 \qquad (8.3.34)$$
$$= \rho\sigma_0 + iZ.$$

We obtain the matrix representation (8.3.10) of a real quaternion and we have seen that the quaternion algebra can be identified with the subalgebra $\boldsymbol{C}_+(E_3)$ of the Clifford algebra $\boldsymbol{C}(E_3)$.

Conversely, every nonzero element h of $\boldsymbol{C}_+(E_3)$ may be written

$$h = \zeta\sigma_0 + iW,$$

where ζ and W are not both zero. As the equations

$$\zeta = \boldsymbol{x}\cdot\boldsymbol{y} \qquad \text{and} \qquad w = \boldsymbol{x} \times \boldsymbol{y}$$

possess an infinite number of solutions, h can always be written in the form $h = XY$.

The nonzero quaternions thus form a multiplicative group denoted $\boldsymbol{G}_+(E_3)$, and it is an example of an EVEN CLIFFORD GROUP. We shall say that this multiplicative group is generated in the algebra $\boldsymbol{C}_+(E_3)$ by the product of vectors of the Euclidean space E_3.

Let us return to the matrix representation (8.3.34) of quaternions. Let us set

$$u = \rho + iz_3 \qquad \text{and} \qquad v = z_2 + iz_1.$$

The matrix (8.3.34) is then written

$$h = \begin{bmatrix} \rho + iz_3 & z_2 + iz_1 \\ -z_2 + iz_1 & \rho - iz_3 \end{bmatrix}$$
$$= \begin{bmatrix} u & v \\ -v^* & u^* \end{bmatrix}. \qquad (8.3.35)$$

In the matrix representation of quaternions the norm $N(h)$ is given by

$$N(h) = \det h = uu^* + vv^*. \qquad (8.3.36)$$

Let us consider, for the moment, the quaternions with norm unity. The subgroup of the Clifford group $\boldsymbol{G}_+(E_3)$ formed by the quaternions with norm unity is called a SPINOR GROUP. It is the spinor group of three-dimensional Euclidean space, and it is denoted $\boldsymbol{Spin}(3)$.

Expression (8.3.35) for h shows that the real quaternion of unit norm can be identified with the matrices of the group $\boldsymbol{SU}(2)$. Thus the spinor group $\boldsymbol{Spin}(3)$ is identified with the unimodular unitary group $\boldsymbol{SU}(2)$.

Minkowski Space

We have seen that the Clifford algebra $\boldsymbol{C}(E_{1,3})$ of Minkowski space can be realized by using the Dirac matrices (8.3.7). It is also possible to realize $C(E_{1,3})$ from the quaternions.

Let us use the matrices $\boldsymbol{e}_1, \boldsymbol{e}_2, \boldsymbol{e}_3$, (8.3.11) as well as the unit matrix $\mathbb{1}$ to form the following super-matrices

$$
\begin{aligned}
\alpha_i &= \begin{bmatrix} 0 & \boldsymbol{e}_i \\ \boldsymbol{e}_i & 0 \end{bmatrix}, \quad i = 1, 2, 3, \\
\alpha_4 &= \begin{bmatrix} 0 & \mathbb{1} \\ -\mathbb{1} & 0 \end{bmatrix}.
\end{aligned} \tag{8.3.37}
$$

The matrices $\alpha_1, \alpha_2, \alpha_3, \alpha_4$ form the generating elements of a sixteen-dimensional Clifford algebra (retaining $(\alpha_i)^2 = -1$ in its definition). It is the algebra $\boldsymbol{H}(2)$ of quaternion matrices which is the Clifford algebra of Minkowski space which is isomorphic to the algebra $\boldsymbol{C}(E_{1,3})$ obtained from the Dirac matrices.

Let us reconsider expression (8.3.12) giving the matrix representation of a quaternion

$$h = a\mathbb{1} + b\boldsymbol{e}_1 + c\boldsymbol{e}_2 + d\boldsymbol{e}_3. \tag{8.3.38}$$

If a, b, c are complex numbers we say that h is a *complex quaternion*. The expression for the matrices $\boldsymbol{e}_j$ given by (8.3.11) allows us to write this quaternion in the form

$$
\begin{aligned}
h &= \begin{bmatrix} a + id & b + ic \\ -b + ic & a - id \end{bmatrix} \\
&= \begin{bmatrix} \alpha & \beta \\ \gamma & \delta \end{bmatrix},
\end{aligned} \tag{8.3.39}
$$

where $\alpha, \beta, \gamma, \delta$ are complex numbers. The norm of a complex quaternion is given by

$$
\begin{aligned}
N(h) &= a^2 + b^2 + c^2 + d^2 \\
&= \det(h).
\end{aligned} \tag{8.3.40}
$$

Let us consider, for the moment, the complex quaternion of unit norm. These latter form a group since

$$
\begin{aligned}
N(hh') &= N(h)N(h') = 1, \\
N(h^{-1}) &= \frac{1}{N(h)} = 1.
\end{aligned} \tag{8.3.41}
$$

The matrices (8.3.39), with determinant equal to unity, are the matrices of the group $\boldsymbol{SL}(2,\mathbb{C})$. In fact,

$$\det h = 1$$

gives us

$$\alpha\delta = \gamma\beta = 1,$$

which characterizes the matrices of the group $\boldsymbol{SL}(2,\mathbb{C})$. The complex quaternions with unit norm thus form a group identical to the unimodular group $\boldsymbol{SL}(2,\mathbb{C})$. The latter is also identified with the spinor group called $\boldsymbol{Spin}(1,3)$.

8.4 Solved Problems

EXERCISE **8.1**

1. Show that the vector space $\mathbb{R}^3$ forms an algebra when provided with the vector product.
2. Show that it is a Lie algebra.
3. Determine the structure constants of this algebra.

SOLUTION **8.1**

1. Let $\boldsymbol{x}$ and $\boldsymbol{y}$ be two vectors of $\mathbb{R}^3$. The vector product

$$\boldsymbol{z} = \boldsymbol{x} \times \boldsymbol{y}$$

is a bilinear product because the properties of the vector product give

$$\left(\sum_i \alpha_i \boldsymbol{x}_i\right) * \left(\sum_j \beta_j \boldsymbol{y}_j\right) = \sum_i \sum_j \alpha_i \beta_j (\boldsymbol{x}_i * \boldsymbol{y}_j).$$

2. Let us consider the Lie bracket defined by the vector product, that is,

$$[\boldsymbol{x}, \boldsymbol{y}] = \boldsymbol{x} \times \boldsymbol{y}.$$

This product satisfies the axioms of the Lie bracket:

— the element $\boldsymbol{z} = \boldsymbol{x} \times \boldsymbol{y}$ is an element of $\mathbb{R}^3$;

— the vector product is distributive and anticommutative

$$\begin{aligned}\boldsymbol{x} \times (\alpha\boldsymbol{y} + \beta\boldsymbol{z}) &= \alpha(\boldsymbol{x} \times \boldsymbol{y}) + \beta(\boldsymbol{x} \times \boldsymbol{z}),\\ \boldsymbol{x} \times \boldsymbol{y} &= -\boldsymbol{y} \times \boldsymbol{x}.\end{aligned}$$

— the following property of the double vector product holds

$$\boldsymbol{a} \times (\boldsymbol{b} \times \boldsymbol{c}) = (\boldsymbol{a}\cdot\boldsymbol{c})\boldsymbol{b} - (\boldsymbol{a}\cdot\boldsymbol{b})\boldsymbol{c}.$$

The Jacobi identity may be written in the form of a double vector product and the use of the previous property gives

$$[\boldsymbol{x}, [\boldsymbol{y}, \boldsymbol{z}]] + [\boldsymbol{y}, [\boldsymbol{z}, \boldsymbol{x}]] + [\boldsymbol{z}, [\boldsymbol{x}, \boldsymbol{y}]] = 0.$$

3. Let $\{\boldsymbol{e}_1, \boldsymbol{e}_2, \boldsymbol{e}_3\}$ be a standard orthonormal basis of $\mathbb{R}^3$. The vector product of the basis elements gives us

$$[\boldsymbol{e}_j, \boldsymbol{e}_k] = \boldsymbol{e}_j \times \boldsymbol{e}_k = \varepsilon_{jkm}\boldsymbol{e}_m, \qquad j, k, m, = 1, 2, 3.$$

The components ε_{jkm} are thus the structure constants of the Lie algebra. For example, for $j = 1$ and $k = 2$ we have

$$\varepsilon_{121} = \varepsilon_{122} = 0, \qquad \varepsilon_{123} = 1.$$

Exercise **8.2**
Consider the operators

$$Q = x \qquad \text{and} \qquad P = \frac{\partial}{\partial x}$$

acting on functions of the real variable x.

1. Calculate the commutator $[P, Q]$.
2. Considering the commutator as the Lie bracket, show that the vector space with basis $\{1, Q, P\}$ is a Lie algebra.
3. Set
$$a = \frac{P + Q}{\sqrt{2}}, \qquad a^\dagger = \frac{Q - P}{\sqrt{2}}.$$
Calculate the commutator $[a, a^\dagger]$.
4. Consider a set of operators Q_i and P_i operating on some independent variables x_i. Find the commutation relations for the operators a_i and ${a_i}^\dagger$ associated with each variable.
5. Consider two operators a_1 and a_2. Show that the operators
$$J_+ = {a_1}^\dagger a_2, \qquad J_- = {a_2}^\dagger a_1, \qquad J_0 = \tfrac{1}{2}({a_1}^\dagger a_1 - {a_2}^\dagger a_2)$$
form a realization of the Lie algebra $\mathfrak{su}(2)$.

SOLUTION **8.2**

1. The commutator applied to a function $f(x)$ gives us

$$[P, Q]f(x) = \frac{\partial}{\partial x} xf(x) - x\frac{\partial}{\partial x} f(x) = f(x),$$

whence

$$[P, Q] = 1.$$

2. The commutator is distributive and anticommutative. On the other hand, the Jacobi identity may be written

$$[1, [P, Q] + [P, [Q, 1]] + [Q, [1, P]] = [1, 1] + [P, 0] + [Q, 0] = 0.$$

The Lie algebra $\{1, Q, P\}$ is called the HEISENBERG ALGEBRA, and it plays a fundamental role in quantum mechanics.

3. The commutator $[a, a^\dagger]$ gives us

$$\begin{aligned} [a, a^\dagger] &= \left(\frac{P+Q}{\sqrt{2}}\right)\left(\frac{Q-P}{\sqrt{2}}\right) - \left(\frac{Q-P}{\sqrt{2}}\right)\left(\frac{P+Q}{\sqrt{2}}\right) \\ &= [P, Q] \\ &= 1. \end{aligned}$$

The operators a and $a^\dagger$ are called respectively *boson creation and annihilation operators*. They are introduced naturally during the solution of the Schrödinger equation for the harmonic oscillator, the solutions of which form a representation of the operators a and $a^\dagger$ in Hilbert space.

4. Since the variables x_i are independent we have the following commutation relations

$$[a_i, a_j] = 0, \qquad [{a_i}^\dagger, {a_j}^\dagger] = 0, \qquad [a_i, {a_j}^\dagger] = \delta_{ij}.$$

5. The operators J_+, J_-, and J_0 satisfy the following commutation relations

$$[J_+, J_-] = 2J_0, \qquad [J_0, J_\pm] = \pm J_\pm.$$

We obtain the commutation relations of the infinitesimal matrices of representations of the group $\boldsymbol{SO}(3)$ given by (3.2.7). These relations are identical to those of the Lie algebra $\mathfrak{su}(2)$. The operators J_+, J_-, and J_0 thus form a realization of $\mathfrak{su}(2)$.

EXERCISE **8.3**

1. Show that the following matrices

$$\gamma_i = \begin{bmatrix} \sigma_i & 0 \\ 0 & \sigma_i \end{bmatrix}, \quad i = 1, 2, 3, \qquad \gamma_0 = \begin{bmatrix} 0 & -i\sigma_2 \\ i\sigma_3 & 0 \end{bmatrix}, \tag{8.4.42}$$

where the σ_i are the Pauli matrices, are able to serve as generating elements of a Clifford algebra.

2. Determine the basis elements of this Clifford algebra.

SOLUTION **8.3**

1. The generating elements of a Clifford algebra have to satisfy the conditions (8.3.1), that is to say,

$$\gamma_i\gamma_j + \gamma_j\gamma_i = \{\gamma_i, \gamma_j\} = 2\delta_{ij} \qquad \text{with} \quad i, j = 0, 1, 2, 3.$$

We obtain the following products

$$\begin{aligned} \{\gamma_i, \gamma_j\} &= \begin{bmatrix} \sigma_i & 0 \\ 0 & \sigma_i \end{bmatrix} \begin{bmatrix} \sigma_j & 0 \\ 0 & \sigma_j \end{bmatrix} + \begin{bmatrix} \sigma_j & 0 \\ 0 & \sigma_j \end{bmatrix} \begin{bmatrix} \sigma_i & 0 \\ 0 & \sigma_i \end{bmatrix} \\ &= \begin{bmatrix} \sigma_i\sigma_j & 0 \\ 0 & \sigma_i\sigma_j \end{bmatrix} + \begin{bmatrix} \sigma_j\sigma_i & 0 \\ 0 & \sigma_j\sigma_i \end{bmatrix} \\ &= \begin{bmatrix} \sigma_i\sigma_j + \sigma_j\sigma_i & 0 \\ 0 & \sigma_i\sigma_j + \sigma_j\sigma_i \end{bmatrix}. \end{aligned}$$

Since the Pauli matrices anticommute we have

$$\sigma_i\sigma_j + \sigma_j\sigma_i = 0 \qquad \text{for} \quad i \neq j,$$

the commutators $\{\gamma_i, \gamma_j\}$ are zero for $i, j = 1, 2, 3$ and $i \neq j$. Similarly we obtain

$$\{\gamma_i, \gamma_0\} = 0, \qquad i = 1, 2, 3.$$

On the other hand, we have

$$(\sigma_i)^2 = 1 \qquad \text{for} \quad i = 0, 1, 2, 3.$$

The defining relations of the generating elements are thus satisfied and are able to serve for forming the basis of a Clifford algebra.

2. In order to obtain the basis of this algebra we form the different products of the generating elements with each other in accordance with the rule (??), adding the element unity. We obtain the following elements

$$\begin{aligned} &\mathbb{1}, \ \gamma_0, \ \gamma_1, \ \gamma_2, \ \gamma_3, \\ &\gamma_0\gamma_1, \ \gamma_0\gamma_2, \ \gamma_0\gamma_3, \ \gamma_1\gamma_2, \ \gamma_1\gamma_3, \ \gamma_2\gamma_3, \\ &\gamma_0\gamma_1\gamma_2, \ \gamma_0\gamma_1\gamma_3, \ \gamma_0\gamma_2\gamma_3, \ \gamma_1\gamma_2\gamma_3, \\ &\gamma_0\gamma_1\gamma_2\gamma_3. \end{aligned}$$

which are all different, and $2^4 = 16$ in number.

Exercise **8.4**

Let there be n Hermitian operators α_j satisfying the Clifford algebra relations

$$\alpha_j\alpha_k + \alpha_k\alpha_j = 2\delta_{jk}.$$

Set

$$a_1 = \tfrac{1}{2}(\alpha_1 + i\alpha_2), \qquad a_1{}^\dagger = \tfrac{1}{2}(\alpha_1 - i\alpha_2), \qquad a_2 = \tfrac{1}{2}(\alpha_3 + i\alpha_4), \quad \text{etc.}$$

Calculate the anticommutation relations $\{a_j, a_k\}$.

Solution **8.4**

We obtain the following anticommutation relations by taking account of the Clifford algebras relations, that is,

$$\{a_i, a_j\} = 0, \qquad \{a_i{}^\dagger, a_j{}^\dagger\} = 0, \qquad \{a_i, a_j{}^\dagger\} = \delta_{ij}.$$

These anticommutation relations are those realized by the fermion creation and annihilation operators in quantum mechanics. They are the operators associated with particles having half-integer spin. Mathematically, they appear as the artefacts of a Clifford algebra.

Exercise **8.5**

This exercise allows us to introduce the relation between the Pauli spinors and quaternions.

1. Let $1, \boldsymbol{i}, \boldsymbol{j}, \boldsymbol{k}$ be the basis vectors of the quaternion algebra. Recall the properties of these vectors.

2. Quaternions can be written $q = w + x\boldsymbol{i} + y\boldsymbol{j} + z\boldsymbol{k}$. Set $\psi = w + z\boldsymbol{k}$, $\phi = -y + x\boldsymbol{k}$. Show that q can be written in the form

$$q = \psi - \boldsymbol{j}\phi.$$

3. Consider a quaternion $u = \alpha + \beta \boldsymbol{k}$, and write $\tilde{u} = \alpha - \beta \boldsymbol{k}$. Let

$$q_1 = \psi_1 - \boldsymbol{j}\phi_1.$$

Show that the product $q_1 q$ can be put in the form

$$q_1 q = \psi'_1 - \boldsymbol{j}\phi'_1,$$

where the quaternions ψ'_1 and ϕ'_1 are functions of $\psi, \phi, \psi_1, \widetilde{\psi}_1, \tilde{\phi}_1$.

4. Consider the following 'rotation quaternion'

$$q_2 = \cos\tfrac{1}{2}\theta - (\boldsymbol{i}L_1 + \boldsymbol{j}L_2 + \boldsymbol{k}L_3)\sin\tfrac{1}{2}\theta,$$

which can be associated with a rotation in three-dimensional space through an angle θ about an axis along a vector $\boldsymbol{L}$ with components L_1, L_2, L_3. Let

$$q_2 q = \psi'_2 - \boldsymbol{j}\phi'_2.$$

Calculate ψ'_2 and ϕ'_2 as a function of the rotation parameters.

5. In the expressions obtained in the preceding exercise replace $\boldsymbol{k}$ by $\sqrt{-1} = i$, and show that the 'vector'

$$\eta = \begin{bmatrix} \psi \\ \phi \end{bmatrix}$$

transforms under a rotation given by q_2 as a two-component spinor.

SOLUTION **8.5**

1. The properties of the basis quaternions are given by formulas (8.3.8), that is,

$$\begin{aligned} \boldsymbol{i}^2 &= \boldsymbol{j}^2 = \boldsymbol{k}^2 = -1, & \boldsymbol{ij} &= -\boldsymbol{ji} = -\boldsymbol{k}, \\ \boldsymbol{jk} &= -\boldsymbol{kj} = -\boldsymbol{i}, & \boldsymbol{ki} &= -\boldsymbol{ij} = -\boldsymbol{j}. \end{aligned}$$

2. The property $\boldsymbol{i} = -\boldsymbol{jk}$ allows q to be written in the form

$$\begin{aligned} q &= (w + z\boldsymbol{k}) - \boldsymbol{j}(-y + x\boldsymbol{k}) \\ &= \psi - \boldsymbol{j}\phi. \end{aligned}$$

3. We obtain the following product

$$\begin{aligned} q_1 q &= (\psi_1 - \boldsymbol{j}\phi_1)(\psi - \boldsymbol{j}\phi) \\ &= \psi_1\psi - \psi_1\boldsymbol{j}\phi - \boldsymbol{j}\phi_1\psi + \boldsymbol{j}\phi_1\boldsymbol{j}\phi. \end{aligned}$$

Using the expressions for the decompositions of $\psi, \phi, \psi_1, \phi_1$ over the basis vectors, each of the terms which appear in the preceding expression can be expanded. For example, we have

$$\begin{aligned}
\boldsymbol{j}\phi_1\boldsymbol{j}\phi &= \boldsymbol{j}(-y_1 + x_1\boldsymbol{k})\boldsymbol{j}\phi \\
&= \boldsymbol{j}[-\boldsymbol{j}(y_1 + x_1\boldsymbol{k})]\phi \\
&= \boldsymbol{j}[-\boldsymbol{j}(-\tilde{\phi}_1)]\phi \\
&= -\tilde{\phi}_1\phi.
\end{aligned}$$

We obtain

$$\begin{aligned}
q_1 q &= (\psi_1\psi - \tilde{\phi}_1\phi) - \boldsymbol{j}(\phi_1\psi + \widetilde{\psi}_1\phi) \\
&= \psi'_1 - \boldsymbol{j}\phi'_1.
\end{aligned}$$

4. The expression for q_2 gives

$$\begin{aligned}
\psi_2 &= \cos\tfrac{1}{2}\theta - \boldsymbol{k}L_3 \sin\tfrac{1}{2}\theta, \\
\phi_2 &= (L_2 - \boldsymbol{k}L_1)\sin\tfrac{1}{2}\theta.
\end{aligned}$$

Replacing q_1 by q_2 in the expression for the product $q_1 q$ obtained in the third question yields

$$\begin{aligned}
\psi'_2 &= \left(\cos\tfrac{1}{2}\theta - \boldsymbol{k}L_3\sin\tfrac{1}{2}\theta\right)\psi - \left((L_2 + \boldsymbol{k}L_1)\sin\tfrac{1}{2}\theta\right)\phi, \\
\phi'_2 &= \left((L_2 - \boldsymbol{k}L_1)\sin\tfrac{1}{2}\theta\right)\psi + \left(\cos\tfrac{1}{2}\theta + \boldsymbol{k}L_3\sin\tfrac{1}{2}\theta\right)\phi.
\end{aligned}$$

5. On replacing the quaternion $\boldsymbol{k}$ by the complex number $i = \sqrt{-1}$ the preceding equations are written

$$\begin{aligned}
\psi'_2 &= \left(\cos\tfrac{1}{2}\theta - iL_3\sin\tfrac{1}{2}\theta\right)\psi - \left((L_2 + iL_1)\sin\tfrac{1}{2}\theta\right)\phi, \\
\phi'_2 &= \left((L_2 - iL_1)\sin\tfrac{1}{2}\theta\right)\psi + \left(\cos\tfrac{1}{2}\theta + iL_3\sin\tfrac{1}{2}\theta\right)\phi.
\end{aligned}$$

We recover formulas (1.2.17) for the transformation of the components (ψ, ϕ) of a spinor under a rotation through an angle θ about an axis along the vector $\boldsymbol{L}$. Quaternions could thus be used in the place of the Pauli spinors—as well as bi-quaternions in the place of Dirac bispinors—which would allow the development of the geometrization of Quantum Mechanics [2].

Appendix
Groups and Their Representations

A.1 The Definition of a Group

Here we recall the principal definitions which we need to know in order to understand the uses of the mathematical notion of group. We refer to our work [11] dedicated to the theory of groups for a deeper study of the applications of this theory in quantum mechanics.

A.1.1 Examples of Groups

Permutation Groups

Let us consider a set of four objects which are numbered from 1 to 4. Let us denote by P the following permutation

$$P = \begin{pmatrix} 1 & 2 & 3 & 4 \\ 2 & 4 & 1 & 3 \end{pmatrix} \tag{A.1.1}$$

which changes the set $(1, 2, 3, 4)$ into the set $(2, 4, 1, 3)$. The permutation P is the *operation* which changes the first set into the second. For example, we can write

$$P(1, 2, 3, 4) = (2, 4, 1, 3). \tag{A.1.2}$$

The number of permutations in this set is equal to $4! = 24$. The *neutral element* E is the identity permutation which leaves any of these sets unchanged. For example,

$$E(2,3,4,1) = (2,3,4,1). \tag{A.1.3}$$

The *inverse* P^{-1} *of a permutation* P defined by (A.1.1) is given by

$$P^{-1}(2,4,1,3) = (1,2,3,4). \tag{A.1.4}$$

Let P_1 be the permutation

$$P_1(1,2,3,4) = (2,3,1,4)$$

and P_2 the permutation

$$P_2(2,3,1,4) = (4,1,2,3).$$

The *product* of these two permutations, denoted

$$P_3 = P_2 P_1,$$

is defined as the permutation which changes $(1,2,3,4)$ into $(4,1,2,3)$, that is to say,

$$P_3(1,2,3,4) = (4,1,2,3). \tag{A.1.5}$$

We can write

$$\begin{aligned} P_3(1,2,3,4) &= P_2[P_1(1,2,3,4)] \\ &= P_2(2,3,1,4) \\ &= (4,1,2,3). \end{aligned} \tag{A.1.6}$$

The permutation operations P_i form the elements of a set which has the properties of a *group*. It is a group which has a finite number of elements, and it constitutes what is called a *finite group*. This group relates to the permutation of four objects, and is denoted S_4.

The preceding properties can be generalized to every set of n objects numbered from 1 to n. The permutations of these n objects form a group which contains $n!$ permutation operations. The group is called the *group of permutations*, or the *symmetric group*. It is denoted as S_n.

Pointed Finite Symmetry Groups

The study of geometric transformations in three-dimensional space, applied to atoms and molecules, is fundamental to quantum physics. Let us study some examples of these transformations.

Let us consider, for example, the case of the water molecule H_2O, which is a plane molecule the three atoms of which are situated at the vertices of

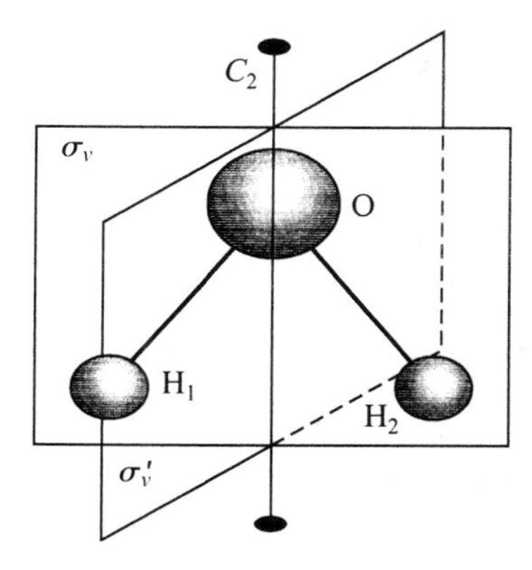

FIGURE B.1. The symmetries of the water molecule.

an isosceles triangle in a plane denoted σ_ν (Figure B.1). The 'symmetry' of a molecule determined by the set of displacements make it coincide with itself. In quantum mechanics these displacements are called *symmetry transformations* or sometimes *symmetry operations*.

A rotation through an angle of π about the molecule's symmetry axis, denoted C_2, takes the hydrogen atom 1 to the location of the hydrogen atom 2. Since the hydrogen atoms are indistinguishable this rotation brings the atom into coincidence with itself. A rotation through an angle π is an example of a *symmetry transformation*.

The molecule H_2O also has two other symmetry elements that are plane:

— first, the plane σ_ν of the molecule, passing through the centers of the three atoms of the molecule;

— second, a plane denoted σ_ν' that is perpendicular to σ_ν and passes through the symmetry axis C_2.

The set of all the symmetry transformations of a given molecule always forms a group, and it is called its group of symmetry transformations, or, more briefly, its *symmetry group*. The symmetry operations therefore make up the elements of the symmetry group.

We have just seen that the water molecule has as symmetry transformations:

— rotations through π and 2π about the axis C_2;

— reflections with respect to the planes σ_ν and σ_ν'.

The group of symmetry transformations of H_2O thus consists of the four following elements

$$C_1; \quad {C_2}^2 = E; \quad \sigma_\nu; \quad \sigma_\nu'. \tag{A.1.7}$$

Such a group is denoted $\boldsymbol{C}_{2\nu}$. Let us note that as every molecule is transformed into itself by the identity operation E, the latter appears in all the groups.

Continuous Symmetry Point Groups

The continuous symmetry point groups have an infinite number of elements that are a function of one or several continuous parameters.

Let us consider, for example, the case of a molecule made up of two different atoms. This molecule possesses a symmetry axis formed by the link passing through the center of the atoms. A rotation through an arbitrary angle φ about this axis makes the molecule coincide with itself. The molecule has an infinity of possible angles of rotation about this axis denoted C_∞. Furthermore, such a molecule possesses an infinite number of reflection planes σ_ν passing through the axis C_∞. The rotations about this axis and the reflections σ_ν make up the elements of a group denoted $\boldsymbol{C}_{\infty\nu}$.

The group denoted $\boldsymbol{O}(3)$ is the symmetry group of an isolated atom. All the rotations about an arbitrary axis passing through the center of the atom, as well as all the reflections in any plane passing through this same center, constitute the elements of the group $\boldsymbol{O}(3)$.

A.1.2 The Axioms Defining a Group

We have seen that the elements of a group can be 'multiplied' with each other. The product of two or more transformations gives a transformation the resultant of which is equivalent to that of their successive application. From this it follows that the product of two or several elements of a group is again an element of this group. It is said that the multiplication of two elements with each other is an INTERNAL COMPOSITION LAW.

Quite generally, in order that a set of elements A, B, C, etc., forms a mathematical group, denoted $\boldsymbol{G}$, it is necessary that the following rules be satisfied:

(G_1) The product of two arbitrary elements A and B of the group must be an element C of the group. The group must thus be provided with an internal composition law, that is to say,

$$AB = C = \text{an element of } \boldsymbol{G}. \qquad \text{(A.1.8)}$$

(G_2) The multiplication of the elements of the group must be associative

$$A(BC) = (AB)C. \qquad \text{(A.1.9)}$$

(G_3) There must exist an element of the group that commutes with all the others and leaves them unchanged. This element is denoted by the letter E; it is the IDENTITY ELEMENT or NEUTRAL ELEMENT. For every arbitrary element A or $\boldsymbol{G}$ we thus have

$$EA = AE = A. \tag{A.1.10}$$

(G_4) Each element A must have an INVERSE ELEMENT which also belongs to $\boldsymbol{G}$. An element, denoted A^{-1}, is called the inverse of A if

$$A^{-1}A = AA^{-1} = E. \tag{A.1.11}$$

A.1.3 Elementary Properties of Groups

The number of elements of a finite group is called its ORDER. The product of an arbitrary element A of a group $\boldsymbol{G}$ with itself must evidently be an element of the group, that is to say,

$$AA = \text{an element of } G.$$

The product AA can be called the *square* of A, and is denoted A^2.

The Neutral Element

The neutral element of a group is unique. In fact, if E and E' both satisfy the definition G_3, then

$$E = E'E = EE' = E'. \tag{A.1.12}$$

The Inverse

The inverse A^{-1} of an element A is determined uniquely. In fact, if B is an inverse of A, by definition G_4 we have

$$BA = AB = E,$$

whence

$$\begin{aligned} B = EB &= (A^{-1}A)B \\ &= A^{-1}(AB) = A^{-1}E \\ &= A^{-1}. \end{aligned} \tag{A.1.13}$$

Thus we have the following theorem about the inverse of a product:

THEOREM. *The inverse of a product of two or more elements is equal to the product of the inverses taken in the reverse order.*

If we consider, for example, three elements A, B, C the theorem means that for the inverse we have

$$(ABC)^{-1} = C^{-1}B^{-1}A^{-1}. \tag{A.1.14}$$

Proof. The proof of this theorem is immediate. Using the associativity property gives

$$\begin{aligned}(ABC)(C^{-1}B^{-1}A^{-1}) &= AB(CC^{-1})B^{-1}A^{-1} \\ &= AB(E)B^{-1}A^{-1} \\ &= AEA^{-1} = E. \end{aligned} \tag{A.1.15}$$

The element $(C^{-1}B^{-1}A^{-1})$ is certainly the inverse of (ABC) because the product of these two elements is the neutral element E. The relation (A.1.14) is thus satisfied. ■

Commutativity

In general the multiplication between elements of a group is not commutative

$$AB \neq BA. \tag{A.1.16}$$

If all the elements of a group are commutative such a group is called an ABELIAN GROUP. The groups which are called CYCLIC are particular cases of abelian groups. A cyclic group of order n has all its element in the form $A, A^2, \ldots, A^n$, where A is an arbitrary element of the group.

Subgroups

In the interior of a group $\boldsymbol{G}$ we can, in fact, find various sets of elements of $\boldsymbol{G}$ which form smaller groups; these subsets are called *subgroups*.

Thus the neutral element E is itself a group of order 1. The neutral element evidently belongs to all the subgroups of a given group. Quite generally, an element of a group may belong to different subgroups.

We also obtain a subgroup of a finite symmetry group by taking an element A of this group and iterating it a certain number of times. Let n be the smallest number for which $A^n = E$; it is said that n is the ORDER OF THE ELEMENT A. The set of elements $A, A^2, \ldots, A^n$ forms a subgroup of the initial group; it is a cyclic subgroup.

The Product Group

Let there be two finite groups $\boldsymbol{G}$ and $\boldsymbol{H}$ or respective orders g and h, the elements of which are respectively $G_1, G_2, \ldots, G_g$ and $H_1, H_2, \ldots, H_h$.

Let us assume that all the elements of $\boldsymbol{G}$ are distinct from those of $\boldsymbol{H}$, except for E, and that they *commute* with the latter. Let us then consider all the pairs (G_i, H_j). We obtain a new set of elements gh which can be considered as a group on choosing as the definition of the product of arbitrary pairs (G_i, H_j) and (G_k, H_l) the formula

$$\begin{aligned}(G_i, H_j)(G_k, H_l) &= (G_iG_k, H_jH_l) \\ &= (G_m, H_q).\end{aligned} \tag{A.1.17}$$

This new set is called the *direct product* of $\boldsymbol{G}$ with $\boldsymbol{H}$, and is *denoted* $\boldsymbol{G} \otimes \boldsymbol{H}$; its order is gh.

Isomorphic Groups

Let two groups $\boldsymbol{G}$ and $\boldsymbol{H}$ of the same order be such that to each element G_i of $\boldsymbol{G}$ we can make correspond an element H_i of $\boldsymbol{H}$ by a one-to-one correspondence. Furthermore, this correspondence must preserve the product, that is to say, it is such that if G_i corresponds to H_i and G_j to H_j then to the element G_iG_j there corresponds the element H_iH_j. In this case it is said that the groups $\boldsymbol{G}$ and $\boldsymbol{H}$ are *isomorphic groups*.

Homomorphic Groups

Let us consider a group $\boldsymbol{G}$ to which we can associate with every element of it one and only one element of another group $\boldsymbol{H}$. Against this, to every element of $\boldsymbol{H}$ there can be associated at least one, and thus possibly more, element(s) of $\boldsymbol{G}$. If, moreover, this correspondence preserves the product in the sense indicated for isomorphic groups, then it is said that the two groups are *homomorphic groups*.

A.2 Linear Operators

A.2.1 The Operator Representing an Element of a Group

EXAMPLE. Let us consider an element of a group, for example a rotation R of geometric space. This rotation transforms a point $\boldsymbol{r}(x, y, z)$ of geometric space into another point $\boldsymbol{r}'(x', y', z')$ such that

$$\boldsymbol{r}' = R\boldsymbol{r}. \tag{A.2.1}$$

On the other hand, let us consider a vector space constructed from geometric space and the vectors $\boldsymbol{X}$ of which are transformed by a rotation of geometric space into some other vectors $\boldsymbol{X}'$. Let us denote this transformation as

$$\boldsymbol{X}' = \Gamma(R)\boldsymbol{X}. \tag{A.2.2}$$

The symbol $\Gamma(R)$ which represents this transformation of the vector $\boldsymbol{X}$ will be called the *rotation operator*. Let us insist on the distinction to be made between the geometrical rotation R operating in physical space and the operator $\Gamma(R)$ which acts in the space of vectors.

In fact, if we consider the vectors OM of the geometric space we may have the impression that there is no difference between the rotation R and the corresponding operator $\Gamma(R)$. However, if we have a vector space whose elements are functions, for example, we see clearly the distinction between R, which acts in the geometric space, and $\Gamma(R)$ which acts in the space of functions. ■

DEFINITION. *Quite generally, an operator is a symbol which when acting on a mathematical entity transforms it into another entity of the same nature. To each element* $\boldsymbol{G}$ *of a group we can associate an operator, denoted* $\Gamma(G)$, *which transforms a vector* $\boldsymbol{X}$ *into another vector* $\boldsymbol{X}'$ *when* $\boldsymbol{G}$ *realizes a transformation.* ■

Let us study some examples of operators.

A.2.2 The Operators Acting on the Vectors of Geometric Space

For each transformation $\boldsymbol{G}$ of three-dimensional space we can create an operator $\Gamma(G)$ which represents the transformation properties of $\boldsymbol{G}$ onto one or more vectors of geometric space.

If we consider transformations of n vectors simultaneously they can be considered as belonging to a vector space of dimension $3n$. We will then associate to each symmetry operation $\boldsymbol{G}$ an operator $\Gamma(G)$ acting simultaneously on n vectors or on a vector having $3n$ components.

Let us observe that an operator $\Gamma(G)$ acting on a vector $\boldsymbol{X}$ only formalizes mathematically its real displacements when the symmetry operation $\boldsymbol{G}$ acts on the geometric space. Consequently, if a first symmetry operation A transforms $\boldsymbol{X}$ into $\boldsymbol{X}'$, say,

$$\boldsymbol{X}' = \Gamma(A)\boldsymbol{X},$$

then a second operation B acting on $\boldsymbol{X}'$ gives a vector $\boldsymbol{X}''$, whence we have

$$\begin{aligned} \boldsymbol{X}'' &= \Gamma(B)\boldsymbol{X}' \\ &= \Gamma(B)[\Gamma(A)\boldsymbol{X}]. \end{aligned} \tag{A.2.3}$$

But the operation (BA) transforms $\boldsymbol{X}$ into $\boldsymbol{X}''$, thus,

$$\boldsymbol{X}'' = \Gamma(BA)\boldsymbol{X}.$$

Since the vector $\boldsymbol{X}$ is arbitrary we have the identity

$$\Gamma(B)\Gamma(A) = \Gamma(BA). \tag{A.2.4}$$

A.2.3 The Operators Acting on Wave Functions

If it is easy to define operators which act on vectors of physical space, it is only slightly less difficult to do so for wave functions forming a vector space. Let us consider a symmetry operation G which transforms a point P with coordinates (x, y, z) into another point P' with coordinates (x', y', z'). For example, let $\psi(x, y, z)$ be a wave function of an electron in an atom of a molecule. Under a symmetry transformation G the molecule coincides with itself and at the same time the operation G transports the value of ψ obtaining at the point P to the point P'. As the coordinate system remains fixed we thus obtain a new wave function, denoted ψ', which is such that its value at P' is equal to that of ψ at P, thus,

$$\psi'(x', y', z') = \psi(x, y, z). \tag{A.2.5}$$

Let us denote by $\Gamma(G)$ the operator which, when applied to a function ψ, yields the function ψ', that is to say,

$$\Gamma(G)\psi = \psi'. \tag{A.2.6}$$

According to (A.2.6) we obtain

$$[\Gamma(G)\psi](x', y', z') = \psi(x, y, z). \tag{A.2.7}$$

Let us denote by G^{-1} the inverse transformation of G; G^{-1} gives us

$$G^{-1}(x', y', z') = (x, y, z). \tag{A.2.8}$$

Substituting the expression for (x, y, z) given by (A.2.8) into $\psi(x, y, z)$ appearing in (A.2.7) we obtain

$$[\Gamma(G)\psi](x', y', z') = \psi[G^{-1}(x', y', z)]. \tag{A.2.9}$$

This latter relation defines the operator $\Gamma(G)$ as allowing us to obtain the function $[\Gamma(G)\psi]$ explicitly.

Arbitrary Configuration Spaces

Let $(x_1, x_2, \ldots, x_n)$ be the coordinates of the particles of a quantum system, that is to say, the coordinates of the configuration space of the wave functions $\psi(x_1, x_2, \ldots, x_n)$. A symmetry operation G transforms the coordinates $(x_1, x_2, \ldots, x_n)$ into $(x'_1, x'_2 \ldots, x'_n)$; let us write it as

$$G(x_1, x_2, \ldots, x_n) = (x'_1, x'_2, \ldots, x'_n).$$

Relation (A.2.7) is generalized by defining the operator associated with G by

$$[\Gamma(G)\psi](x'_1, x'_2, \ldots, x'_n) = \psi(x_1, x_2, \ldots, x_n). \tag{A.2.10}$$

The transformation G^{-1} gives us

$$G^{-1}({x'}_1, {x'}_2, \ldots, {x'}_n) = (x_1, x_2, \ldots, x_n).$$

Relation (A.2.10) is then written

$$[\Gamma(G)\psi]({x'}_1, {x'}_2, \ldots, {x'}_n) = \psi[G^{-1}({x'}_1, {x'}_2, \ldots, {x'}_n)]. \qquad \text{(A.2.11)}$$

Since the two sides of this relation are expressed as functions of the same variables, the notation of which is arbitrary, for an arbitrary point of the configuration space we shall have

$$[\Gamma(G)\psi](x_1, x_2, \ldots, x_n) = \psi[G^{-1}(x_1, x_2, \ldots, x_n)]. \qquad \text{(A.2.12)}$$

Properties of Operators

Let us show that the operators defined by (A.2.12) are linear. Let us write as x the set of n coordinates of a point, and as x' the coordinates of its transform, that is to say,

$$x' = G(x).$$

Let $\psi(x)$ and $\varphi(x)$ be two arbitrary functions. According to (A.2.12) we have

$$\begin{aligned}
[\Gamma(G)(a\psi + b\phi)](x) &= (a\psi + b\phi)[G^{-1}(x)] \\
&= a\psi[G^{-1}(x)] + b\phi[G^{-1}(x)] \\
&= a[\Gamma(G)\psi](x) + b[\Gamma(G)\phi][(x). \qquad \text{(A.2.13)}
\end{aligned}$$

This latter relation shows that the operator $\Gamma(G)$ is linear. Let us show that the operators $\Gamma(G)$ also possess the following property

$$\Gamma(A)\Gamma(B) = \Gamma(AB), \qquad \text{(A.2.14)}$$

where A and B are two arbitrary elements of the given group. For every function $\psi(x)$ we have, in fact, by applying the operators successively

$$\begin{aligned}
[\Gamma(A)][\Gamma(B)\psi]](x) &= [\Gamma(B)\psi](A^{-1}x) \\
&= \psi[B^{-1}(A^{-1}x)] \\
&= \psi[(AB)^{-1}(x)] \\
&= [\Gamma(AB)\psi](x). \qquad \text{(A.2.15)}
\end{aligned}$$

As relation (A.2.15) is satisfied whatever may be the function $\psi(x)$, we therefore have the identity (A.2.14) between operators.

A.2.4 Operators Representing a Group

Just as for vectors of physical space or wave functions, it is known that these mathematical entities can all be considered as the elements of a vector space.

If an arbitrary vector space V_n of finite dimension n is given, every vector $\boldsymbol{x}$ of this space can be decomposed over the basis vectors $\boldsymbol{e}_i$, that is to say,

$$\boldsymbol{x} = x_1\boldsymbol{e}_1 + x_2\boldsymbol{e}_2 + \cdots + x_n\boldsymbol{e}_n. \tag{A.2.16}$$

If to each symmetry transformation G of a group $\boldsymbol{G}$ we can bring into correspondence a linear operator $\Gamma(G)$ which, when applied to a vector $\boldsymbol{x}$, gives a transformed vector

$$\Gamma(G)\boldsymbol{x} = \sum_{i=1}^{n} x_i\Gamma(G)\boldsymbol{e}_i, \tag{A.2.17}$$

then the operator $\Gamma(G)$ will be defined if the transforms $\Gamma(G)\boldsymbol{e}_i$ of the basis vectors are given. Furthermore, in order to respect properties (A.2.4) and (A.2.14) we impose on the operators $\Gamma(G)$ which are chosen later on that they satisfy the relation

$$\Gamma(B)\Gamma(A) = \Gamma(BA), \tag{A.2.18}$$

where A and B are arbitrary elements of the group $\boldsymbol{G}$.

To each element G of a group we can bring into correspondence an operator $\Gamma(G)$ acting on a vector space V_n and satisfying property (A.2.18). It is said that these operators represent the group $\boldsymbol{G}$, or, again, that they form a REPRESENTATION of the group. We are going to see that it is, in fact, the matrices of these operators which in practice will be used as representations of the group.

The vector space V_n of arbitrary dimension n is called the *representation space of the group* $\boldsymbol{G}$.

A.3 Matrix Representations

Every linear operator acting on vectors of a finite-dimensional vector space can be represented by a matrix defined with respect to the basis vectors of the space.

Let us give two examples.

A.3.1 The Rotation Matrix Acting on the Vectors of a Three-Dimensional Space

Let C_α be a rotation through an angle α in three-dimensional space. Let us consider the rotation operator $\Gamma(C_\alpha)$ acting on the orthonormal basis vectors $\{\boldsymbol{e}_1, \boldsymbol{e}_2, \boldsymbol{e}_3\}$ of the vector space E_3 representing physical space. The transforms of the basis vectors after rotation are

$$\begin{aligned} \Gamma(C_\alpha)\boldsymbol{e}_1 &= \cos\alpha \boldsymbol{e}_1 + \sin\alpha \boldsymbol{e}_2, \\ \Gamma(C_\alpha)\boldsymbol{e}_2 &= -\sin\alpha \boldsymbol{e}_1 + \cos\alpha \boldsymbol{e}_2, \\ \Gamma(C_\alpha)\boldsymbol{e}_3 &= \boldsymbol{e}_3. \end{aligned} \tag{A.3.1}$$

Quite generally, as every vector can be decomposed in the basis $\{\boldsymbol{e}_1, \boldsymbol{e}_2, \boldsymbol{e}_3\}$ we can write

$$\begin{aligned} \Gamma(C_\alpha)\boldsymbol{e}_i &= G_{1i}\boldsymbol{e}_1 + G_{2i}\boldsymbol{e}_e + G_{3i}\boldsymbol{e}_3 \\ &= \sum_{j=1}^{3} G_{ji}\boldsymbol{e}_j, \end{aligned} \tag{A.3.2}$$

with $i = 1, 2, 3$. In the present case relations (A.3.1) and (A.3.2) give us

$$\begin{array}{llll} G_{11} = \cos\alpha, & G_{21} = \sin\alpha, & G_{31} = 0, & G_{12} = -\sin\alpha, \\ G_{22} = \cos\alpha, & G_{32} = 0, & G_{13} = G_{23} = 0, & G_{33} = 1. \end{array} \tag{A.3.3}$$

The quantities G_{ij} yield a numerical representation of the operator $\Gamma(C_\alpha)$ with respect to the basis vectors. It is useful to arrange the numbers G_{ij} in a table making up a square matrix of order 3 called the ROTATION MATRIX *of the operator* $\Gamma(C_\alpha)$. The quantities G_{ji} form the elements of the ith COLUMN of the matrix. Let us denote this matrix by $M(C_\alpha)$; we have

$$M(C_\alpha) = \begin{bmatrix} \cos\alpha & -\sin\alpha & 0 \\ \sin\alpha & \cos\alpha & 0 \\ 0 & 0 & 1 \end{bmatrix}. \tag{A.3.4}$$

A.3.2 The Matrix of an Operator Acting on Functions

The wave functions can be considered as vectors of a vector space, and in practice we can often be restricted to a finite-dimensional vector space. The operators acting on the wave functions are thus also going to be represented by matrices defined with respect to the basis vectors of this vector space. Let us look at an example.

The Reflection Matrix

Let us consider the vector space E_2 having as basis the following functions

$$p_x = f(r)\sin\theta\cos\varphi, \qquad p_y = f(r)\sin\theta\sin\varphi, \tag{A.3.5}$$

where the variables r, θ, φ are the spherical coordinates. Let us study the transformations of these functions by a reflection σ_ν in a plane xOz. Only the angle φ is changed into $\varphi' = -\varphi$ by such a reflection, and we have

$$[\Gamma(\sigma_\nu)p_x](r,\theta,\varphi) = f(r)\sin\theta\cos(-\varphi) = p_x, \tag{A.3.6}$$

$$[\Gamma(\sigma_\nu)p_y](r,\theta,\varphi) = f(r)\sin\theta\sin(-\varphi) = -p_y, \tag{A.3.7}$$

whence the reflection matrix

$$M(\sigma_\nu) = \begin{bmatrix} 1 & 0 \\ 0 & -1 \end{bmatrix}. \tag{A.3.8}$$

We note that this matrix is identical to that of an operator $\Gamma(\sigma_\nu)$ transforming basis vectors $\boldsymbol{e}_1$ and $\boldsymbol{e}_2$ of a Cartesian plane of reference Oxy.

A.3.3 The Matrices Representing the Elements of a Group

Quite generally, a linear operator $\Gamma(G)$ acting on the vectors $\boldsymbol{x}$ of a vector space V_n yields the transformed vector

$$\Gamma(G)\boldsymbol{x} = \sum_{i=1}^{n} x_i \Gamma(G)\boldsymbol{e}_i. \tag{A.3.9}$$

The decomposition of the transforms $\Gamma(G)\boldsymbol{e}_i$ of the basis vectors is

$$\Gamma(G)\boldsymbol{e}_i = \sum_{j=1}^{n} G_{ji}\boldsymbol{e}_i. \tag{A.3.10}$$

The numbers G_{ji} form the elements of the matrix of the operator $\Gamma(G)$ with respect to the basis $\{\boldsymbol{e}_i\}$ of the vector space V_n. To each element G of a group $\boldsymbol{G}$ we can thus associate a matrix $M(G) = [G_{ji}]$ such that

$$M(G) = \begin{bmatrix} G_{11} & G_{12} & \dots & G_{1n} \\ G_{21} & G_{22} & \dots & G_{2n} \\ \vdots & \vdots & \vdots & \vdots \\ G_{n1} & G_{n2} & \dots & G_{nn} \end{bmatrix}. \tag{A.3.11}$$

Let us note that the components G_{ji} of the vector $\Gamma(G)\boldsymbol{e}_i$ provide the elements of the ith column of $M(G)$. The matrices $M(G)$ are square matrices of order n equal to the dimension of V_n. We shall say that the matrices $M(G)$ represent the elements G of the symmetry group.

A.4 Matrix Representations

A.4.1 The Definition of a Matrix Representation

Consider a group $\boldsymbol{G}$ and an n-dimensional vector space V_n with basis $\{\boldsymbol{e}_i\}$. Let us bring into correspondence with every element G of the group a linear operator $\Gamma(G)$ acting on the vectors of V_n. If A and B are two elements of $\boldsymbol{G}$ the operators chosen must satisfy the property

$$\Gamma(BA) = \Gamma(B)\Gamma(A). \tag{A.4.1}$$

If the transforms $\Gamma(G)\boldsymbol{e}_i$ of the basis vectors are given we thus define the operator $\Gamma(G)$ and consequently the matrix $M(G)$ of the operator. In fact, the matrix elements $M_{ji}(G)$ of $M(G)$ are the components of the vector $\Gamma(G)\boldsymbol{e}_i$, that is to say,

$$\Gamma(G)\boldsymbol{e}_i = \sum_{j=1}^{n} M_{ji}(G)\boldsymbol{e}_j. \tag{A.4.2}$$

The set of matrices $M(G)$ each representing an element G of the group forms a REPRESENTATION OF THE GROUP IN A MATRIX FORM. The vector space V_n is called the REPRESENTATION SPACE and the basis $\{\boldsymbol{e}_i\}$ is a REPRESENTATION BASIS. If the operators are denoted $\Gamma(G)$ and if n is the dimension of V_n the representation will be denoted, for example, by Γ_n.

REMARK. Since the vector space which is chosen can be arbitrary the representations can be made up of matrices of arbitrary order. Furthermore, for a given vector space the matrices depend upon the basis chosen. We thus have an infinite number of possible representations in which there is no a priori interest. However, for a given group all these representations can be decomposed into a certain number of representations called IRREDUCIBLE REPRESENTATIONS which constitute a well-defined characteristic of a group and allow the applications of group theory in physics. ■

A.4.2 The Fundamental Property of the Matrices of a Representation

Let A and B be elements of a group $\boldsymbol{G}$. Let us show that the matrices $M(G)$ of a representation of $\boldsymbol{G}$ satisfy the relation

$$M(B)M(A) = M(BA). \tag{A.4.3}$$

For that let us apply the operators $\Gamma(A)$ and $\Gamma(B)$ successively to a basis vector $\boldsymbol{e}_i$ of the representation space, that is to say,

$$\Gamma(B)[\Gamma(A)\boldsymbol{e}_k] = \Gamma(B)\sum_{l=1}^{n} M_{lk}(A)\boldsymbol{e}_l$$

$$= \sum_{l,m} M_{lk}(A) M_{ml}(B) e_m$$

$$= \sum_m \left[\sum_l M_{ml}(B) M_{lk}(A) \right] e_M. \qquad \text{(A.4.4)}$$

On the other hand, property (A.2.18) of the definition of the operators of a representation gives us

$$\begin{aligned} \Gamma(B)[\Gamma(A)e_k] &= \Gamma(BA)e_k \\ &= \sum_m M_{mk}(BA) e_m. \end{aligned} \qquad \text{(A.4.5)}$$

The comparison of (A.4.4) and (A.4.5) gives us

$$M_{mk}(BA) = \sum_m M_{ml}(B) M_{lk}(A). \qquad \text{(A.4.6)}$$

The element $M_{mk}(AB)$ of the matrix $M(AB)$ is equal to the element of the product of the matrices $M(B)$ by $M(A)$, which proves property (A.4.3).

A.4.3 Representation by Regular Matrices

A matrix is called a REGULAR MATRIX if its determinant is not zero and in this case is invertible. If its determinant is zero the matrix is called a SINGULAR MATRIX.

We are going to require that the matrices of the representations are necessarily regular. Such a choice in fact allows us to associate the matrix unity I to the identity element E of a group—a choice which seems quite natural.

Let there be an n-dimensional representation the matrix $M(E)$ of which is assumed to be regular. Its inverse $[M(E)]^{-1}$ exists and we can write

$$\begin{aligned} M(E)M(E)[M(E)]^{-1} &= M(E)I_n \\ &= M(E), \end{aligned} \qquad \text{(A.4.7)}$$

where I_n is the unit matrix of order n. On the other hand, by (A.4.3) we have

$$\begin{aligned} M(E)M(E) &= M(EE) \\ &= M(E), \end{aligned}$$

whence

$$\begin{aligned} M(E)M(E)[M(E)]^{-1} &= M(E)[M(E)]^{-1} \\ &= I_n. \end{aligned} \qquad \text{(A.4.8)}$$

Relations (A.4.7) and (A.4.8) thus yield

$$M(E) = I_n, \tag{A.4.9}$$

a result which is deduced simply from the hypothesis of regularity of the matrices of the representation. From relations (A.4.3) and (A.4.9) we deduce that for every element G of the group

$$\begin{aligned} M(G)M(G^{-1}) &= M(GG^{-1}) \\ &= M(E) \\ &= I_n \\ &= M(G)[M(G)]^{-1}. \end{aligned} \tag{A.4.10}$$

Consequently we have

$$M(G^{-1}) = [M(G)]^{-1}. \tag{A.4.11}$$

The matrix representing the inverse G^{-1} of an element G is the inverse matrix of the matrix representing G.

The regular matrices $M(G)$ of a representation form a group. The neutral element of the group is the matrix

$$M(E) = I_n$$

and the inverse of a matrix exists. The internal composition law of the group is the multiplication between the matrices. Finally, when we associate with each element G of a group a regular matrix satisfying the fundamental property (A.4.3) we shall say, by definition, that these matrices form a representation of the group.

We may therefore reserve the following definition of a matrix representation:

DEFINITION. *When a regular matrix is associated with each element G of a group, and if these matrices satisfy the property*

$$M(B)M(A) = M(BA)$$

it will be said that these matrices form a MATRIX REPRESENTATION *of the group.*

A.4.4 Equivalent Representations

Let $\{\mathbf{e}_i\}$ and $\{\mathbf{e}'_j\}$ be two bases of a representation vector space V_n. These bases are related to each other by the relations

$$\mathbf{e}'_i = \sum_j A_{ji}\mathbf{e}_j, \qquad \mathbf{e}_j = \sum_i (A^{-1})_{ij}\mathbf{e}'_i. \tag{A.4.12}$$

Let us study how the matrices Γ_n of a representation transform under a change of basis defined by (A.4.12). Applying the operator $\Gamma(G)$ at e'_i yields

$$\begin{aligned}
\Gamma(G)e'_i &= \sum_j A_{ji}A_{ji}\Gamma(G)e_j \\
&= \sum_{j,l} A_{ji}M_{lj}(G)e_l \\
&= \sum_{j,l,m} A_{ji}M_{lj}(G)(A^{-1})_{ml}e'_m \\
&= \sum_m \left[\sum_{j,l}(A^{-1})_{ml}M_{lj}(G)a_{ji}\right]e'_m \\
&= \sum_m M'_{mi}(G)e'_m,
\end{aligned} \tag{A.4.13}$$

where the quantities $M'_{mi}(G)$ represent the components of the vector $\Gamma(G)e'_i$ in the basis e'_m. Let us denote by A, A^{-1}, $M(G)$ and $M'(G)$ the matrices which have, respectively, as their elements $A_{ij}(A^{-1})_{kl}$, $M_{lj}(G)$, $M'_{mi}(G)$. The penultimate term of (A.4.13) brings to light the elements of the product matrix

$$M'(G) = A^{-1}M(G)A. \tag{A.4.14}$$

The matrices $M'(G)$ are the matrices of a new representation Γ'_n of the group $\boldsymbol{G}$. In fact, for every element G_i and G_j of the group we have

$$\begin{aligned}
M'(G_i)M'(G_j) &= [A^{-1}M(G_i)A][A^{-1}M(G_j)A] \\
&= A^{-1}M(G_i)M(G_j)A \\
&= A^{-1}M(G_iG_j)A \\
&= M'(G_iG_j).
\end{aligned} \tag{A.4.15}$$

The representation Γ_n', given in the matrix form $M'(G)$, is said to be EQUIVALENT (or *isomorphic*) to the representation Γ_n formed by the matrices $M(G)$. As these two representations are deduced from each other by a change of basis vectors are not essentially different, and they will be able to be identified with each other according to the requirements. The matrices of two equivalent representations are called *equivalent matrices*; they are characterized by property (A.4.14).

A representation given in the form of unitary matrices is called a UNITARY REPRESENTATION. It will be proved that for finite groups every representation of a group is equivalent to a unitary representation of this group.

A.5 Reducible and Irreducible Representations

A.5.1 *The Direct Sum of Two Vector Spaces*

Let V_n be a vector space and let W_m and $W_p{}'$ be two subspaces of V_n. If every vector $\boldsymbol{x}$ of V_n is written uniquely in the form

$$\boldsymbol{x} = \boldsymbol{w} + \boldsymbol{w}',$$

with $\boldsymbol{w}$ and $\boldsymbol{w}'$ belonging, respectively, to W_m and $W_p{}'$, it is then said that V_n is the DIRECT SUM of W_m and $W_p{}'$, and we write

$$V_n = W_m \oplus W_p{}'. \tag{A.5.1}$$

The dimension n of V_n is evidently equal to the sum of the dimensions m and p of W_m and $W_p{}'$. The intersection of W_m and W_p is reduced to zero. It is said that W_p is the COMPLEMENT of W_m in V_n.

EXAMPLE. Let us consider the vector space E_3 of physical space having the vectors $\boldsymbol{e}_1, \boldsymbol{e}_2, \boldsymbol{e}_3$ as basis. As these are linearly independent, every pair $(\boldsymbol{e}_i, \boldsymbol{e}_j)$ constitutes a basis of a two-dimensional vector subspace E_2. In geometric terms E_2 is a plane. Let us, for example, choose $\boldsymbol{e}_1$ and $\boldsymbol{e}_2$ to form a vector space E_2, and let E_1 be the space with $\boldsymbol{e}_1$ as its basis. We have

$$E_3 = E_1 \oplus E_2. \blacksquare \tag{A.5.2}$$

Orthogonal Subspaces

If V_n is provided with a Hermitian product $\langle \boldsymbol{x}, \boldsymbol{y} \rangle$, two vectors $\boldsymbol{x}$ and $\boldsymbol{y}$ are said to be *mutually orthogonal* if

$$\langle \boldsymbol{x}, \boldsymbol{y} \rangle = 0.$$

The set of elements of V_n which are orthogonal to all the elements of a subspace W_1 of V_n form a subspace W_2 of V_n called the SUBSPACE ORTHOGONAL *to* W_1. The subspaces W_1 and W_2 are complementary and V_n is the direct sum $W_1 \oplus W_2$.

A.5.2 *The Direct Sum of Two Representations*

Let us consider a group $\boldsymbol{G}$ and a vector space V_n provided with a Hermitian product $\langle \boldsymbol{x}, \boldsymbol{y} \rangle$. Let us assume that the Hermitian product is invariant under $\boldsymbol{G}$, in other words, for every element G of the group we have

$$\langle \boldsymbol{x}, \boldsymbol{y} \rangle = \langle \Gamma(G)\boldsymbol{x}, \Gamma(G)\boldsymbol{y} \rangle, \tag{A.5.3}$$

which is the case for unitary operators $\Gamma(G)$.

Let W_1 and W_2 be two orthogonal subspaces of V_n and let us denote, respectively, as $\boldsymbol{x}_1$ and $\boldsymbol{x}_2$ the vectors of these subspaces. By hypothesis we have

$$\begin{aligned} \langle \boldsymbol{x}_1, \boldsymbol{x}_2 \rangle &= \langle \Gamma(G)\boldsymbol{x}_1, \Gamma(G)\boldsymbol{x}_2 \rangle \\ &= 0. \end{aligned} \tag{A.5.4}$$

Since every vector $\Gamma(G)\boldsymbol{x}_1$ belongs to W_1 relation (A.5.4) shows that every vector $\Gamma(G)\boldsymbol{x}_2$ belongs to W_2, therefore that W_2 is stable under $\boldsymbol{G}$. We therefore end up with the following theorem:

THEOREM. *Let Γ be a unitary representation of a group $\boldsymbol{G}$, let the representation space V_n be provided with a Hermitian product, and let W_1 be a vector subspace of V_n which is stable under $\boldsymbol{G}$. Then there exists a complement, W_2, of W_1 in V_n which is stable under $\boldsymbol{G}$. W_2 is the subspace orthogonal to W_1.* ■

Each of the subspaces W_1 and W_2 are able to serve as a representation space of a group $\boldsymbol{G}$. Let Γ_1 be the representation defined on W_1 and Γ_2 the representation on W_2. We can also form a representation Γ by taking the space

$$V_n = W_1 \oplus W_2$$

as the representation space. It is said that the representation Γ is the DIRECT SUM *of the representations* Γ_1 and Γ_2, and is written

$$\Gamma = \Gamma_1 \oplus \Gamma_2. \tag{A.5.5}$$

A.5.3 Irreducible Representations

If Γ is a linear representation of a group $\boldsymbol{G}$ which is defined on a vector space V_n it is said that Γ is an IRREDUCIBLE REPRESENTATION if no vector subspace of V_n is stable under $\boldsymbol{G}$. In the opposite case, it is called REDUCIBLE.

In other words, if Γ is an irreducible representation it cannot be decomposed into the direct sum of two representations. Against this, irreducible representations are going to serve for constructing the others by direct sums. We have the following theorem:

THEOREM. *Every representation Γ is the direct sum of irreducible representations.* ■

Proof. In fact, if Γ is itself irreducible there is nothing to prove. Otherwise we can decompose Γ into the direct sum $\Gamma_1 \oplus \Gamma_2$. If these representations Γ_i are themselves reducible we carry on with their decomposition into direct sums until irreducible representations are obtained. A particular representation Γ_i may possibly appear several times in a reducible representation. Finally a decomposition will be obtained in the form

$$\Gamma = \Gamma_1 \oplus \Gamma_2 \oplus \cdots \oplus \Gamma_p. \tag{A.5.6}$$

Each symmetry group is characterized by certain types of irreducible representations Γ_i which are listed in the CHARACTER TABLES of the group. These tables allow us easily to determine the number of times that an irreducible representation Γ_i occurs in a given reducible representation.

A.6 The Direct Product of Representations

A.6.1 The Direct Product of Two Matrices

Consider two vector spaces V and V' or respective dimensions n and n'. Let us denote by $\boldsymbol{x}_i$ and $\boldsymbol{x}'_j$ the respective basis vectors of the spaces V and V'. The tensor product [10] of the vector spaces V and V' is a vector space of dimension nn', denoted $V \otimes V'$, having as basis the tensors

$$\boldsymbol{v}_{ij} = \boldsymbol{x}_i \otimes \boldsymbol{x}'_j. \tag{A.6.1}$$

EXAMPLE. Let us consider two systems $\{\Psi_i\}$ and $\{\Psi'_j\}$ of wave functions, each respectively forming a basis of the vector spaces V and V'. The products $\Psi_i\Psi'_j$ yield a new system of functions which are able to serve as the basis of a vector space which is the tensor product of the spaces V and V'. ■

The Direct Product of the Matrices of Two Representations

Let us consider a group $\boldsymbol{G}$ two representations, Γ and Γ', of which are given with respective representation spaces V and V'. The matrices $M(G)$ and $M'(G)$ of these representations are those of the operators $\Gamma(G)$ and $\Gamma'(G)$ such that

$$\begin{aligned} \Gamma(G)\boldsymbol{x}_i &= \sum_k M_{ki}(G)\boldsymbol{x}_k, \\ \Gamma'(G)\boldsymbol{x}'_j &= \sum_l M'_{lj}(G)\boldsymbol{x}'_l. \end{aligned} \tag{A.6.2}$$

Let us seek a representation of $\boldsymbol{G}$ defined on the space $V \otimes V'$. To do so let us consider the operators $\Gamma''(G)$ defined by

$$\begin{aligned} \Gamma''(G)\boldsymbol{v}_{ij} &= [\Gamma(G)\boldsymbol{x}_i] \otimes [\Gamma'(G)\boldsymbol{x}'_j] \\ &= \left(\sum_k M_{ki}(G)\boldsymbol{x}_k\right) \otimes \left(\sum_l M'_{lj}(G)\boldsymbol{x}'_l\right). \end{aligned} \tag{A.6.3}$$

Let us expand the latter expression by using the linearity properties of the tensor product. This yields

$$\begin{aligned}\Gamma''(G)\boldsymbol{v}_{ij} &= \sum_{kl} M_{ki}(G)M'_{lj}(G)\boldsymbol{x}_k \otimes \boldsymbol{x}'_l \\ &= \sum_{kl} M_{ki}(G)M'_{kj}(G)\boldsymbol{v}_{kl} \\ &= \sum_{kl} M''_{ki,lj}(G)\boldsymbol{v}_{kl}. \end{aligned} \qquad \text{(A.6.4)}$$

This equation gives the expression for the elements of the matrix $M''(G)$ defining the operator $\Gamma''(G)$ in the basis $\{\boldsymbol{v}_{ij}\}$ of the vector space $V \otimes V'$. We shall say that the matrix $M''(G)$ is the TENSOR PRODUCT OF THE MATRIX $M(G)$ WITH THE MATRIX $M'(G)$, and is denoted

$$M''(G) = M(G) \otimes M'(G).$$

The matrix $M''(G)$ is again called the *direct product* of the matrices $M(G)$ and $M'(G)$.

DEFINITION. *Quite generally, consider two matrices* $A = [a_{ij}]$ *and* $B = [b_{kl}]$ *of respective orders* (m_1, n_1) *and* (m_2, n_2). *Their tensor product is a matrix* C *denoted* $A \otimes B$, *of order* (m_1m_2, n_1n_2) *defined by*

$$C = A \otimes B = \begin{bmatrix} a_{11}B & a_{12}B & \dots & a_{1n_1}B \\ a_{21}B & a_{22}B & \dots & a_{2n_1}B \\ \vdots & \vdots & \vdots & \vdots \\ a_{m_11}B & a_{m_12}B & \dots & a_{m_1n_1}B \end{bmatrix}. \quad \blacksquare \qquad \text{(A.6.5)}$$

Each submatrix $a_{ij}B$ possesses elements of the form $a_{ij}b_{kl}$, and these constitute the elements $c_{ij,kl}$ of the matrix C, that is to say,

$$c_{ij,kl} = a_{ij}b_{kl}. \qquad \text{(A.6.6)}$$

The disposition of these elements in the matrix C is realized as in the matrix of (A.6.5).

A.6.2 Properties of Tensor Products of Matrices

The Fundamental Property

Let us consider four matrices A, B, C, and D for which we can form the products AB and CD. Let us show that we have the property

$$AB \otimes CD = (A \otimes C)(B \otimes D). \qquad \text{(A.6.7)}$$

To do so let us write the general terms of the matrix $AB \otimes CD$ and expand it. Let us denote by A_{ij} the elements of the matrix A, etc. This yields

$$\begin{aligned}(AB \otimes CD)_{ij,kl} &= (AB)_{ij}(CD)_{kl} \\ &= \sum_{\alpha,\beta}(A_{i\alpha}B_{\alpha j})(C_{k\beta}D_{\beta l}) \\ &= \sum_{\alpha,\beta}(A_{i\alpha}C_{k\beta})(B_{\alpha j}D_{\beta l}) \\ &= \sum_{\alpha,\beta}(A \otimes C)_{i\alpha,k\beta}(B \otimes D)_{\alpha j,\beta l} \\ &= [(A \otimes C)(B \otimes D)]_{ij,kl}. \end{aligned} \tag{A.6.8}$$

We obtain the identity of the general elements of the matrices $(AB) \otimes (CD)$ and $(A \otimes C)(B \otimes D)$, which proves property (A.6.7). ■

The Inverse of a Tensor Product of Matrices

Using (A.6.7), in the case where the matrices A and C are invertible we have the relation

$$\begin{aligned}(A \otimes C)(A^{-1} \otimes C^{-1}) &= (AA^{-1}) \otimes (CC^{-1}) \\ &= I \otimes I \\ &= I. \end{aligned} \tag{A.6.9}$$

As a consequence $(A^{-1} \otimes C^{-1})$ is the inverse matrix of $(A \otimes C)$, that is to say,

$$(A \otimes C)^{-1} = A^{-1} \otimes C^{-1}. \tag{A.6.10}$$

Unitary Matrices

If two matrices A and B are unitary then so is their tensor product. To prove it let us use the relation between adjoint matrices

$$(A \otimes B)^{\dagger} = A^{\dagger} \otimes B^{\dagger}, \tag{A.6.11}$$

a relation which is evident because of definition (A.6.5) of the tensor product of matrices. As the matrices A and B are unitary we have

$$A^{\dagger} = A^{-1} \qquad \text{and} \qquad B^{\dagger} = B^{-1},$$

whence

$$\begin{aligned}(A \otimes B)^{\dagger} &= A^{-1} \otimes B^{-1} \\ &= (A \otimes B)^{-1}, \end{aligned} \tag{A.6.12}$$

which proves that $(A \otimes B)$ is unitary.

A.6.3 *The Direct Product of Two Representations*

If two representations Γ and Γ' are given by their matrices $M(G)$ and $M'(G)$ we shall show that the matrices $M(G) \otimes M'(G)$ also form a representation. Let two elements G_i and G_j be of a group $\boldsymbol{G}$, then we have

$$\begin{aligned} M(G_i)M(G_j) &= M(G_iG_j), \\ M'(G_i)M'(G_j) &= M'(G_iG_j). \end{aligned} \tag{A.6.13}$$

Relation (A.6.7) then gives us

$$\begin{aligned} &[M(G_i) \otimes M'(G_i)][M(G_j) \otimes M'(G_j)] \\ &\quad = [M(G_i)M(G_j)] \otimes [M'(G_i)M'(G_j)] \\ &\quad = M(G_iG_j) \otimes M'(G_iG_j). \end{aligned} \tag{A.6.14}$$

The matrices

$$M''(G) = M(G) \otimes M'(G)$$

certainly satisfy the properties of a representation Γ'', since (A.6.14) may be written

$$M''(G_i)M''(G_j) = M''(G_iG_j). \tag{A.6.15}$$

The representation Γ'' is given by the matrices $M''(G)$. Γ'' is called the DIRECT PRODUCT OF THE REPRESENTATIONS Γ *and* Γ', and is denoted

$$\Gamma'' = \Gamma \otimes \Gamma'.$$

If V and V' are respectively representation spaces of Γ and Γ; then the tensor product $V \otimes V'$ is the representation space of $\Gamma \otimes \Gamma'$.

References

[1] Benn, I.M. and Tucker, R.W. (1989) *An Introduction to Spinors and Geometry with Applications in Physics* (Adam Hilger: New York)

[2] Boudet, R. (1997) *The Present Status of the Quantum Theory of Light* (eds. S. Jeffers et al.) (Kluwer Academic Publishers: Dordrecht), pp. 471–481

[3] Cartan, E. (1938) *Leçons sur la théorie des spineurs.* Tome I: *Les spineurs de l'espace à trois dimensions.* Tome II: *Les spineurs de l'espace à* $n > 3$ *dimensions. Les spineurs en géometrie riemannienne* (Hermann: Paris)

[4] Cartan, E. (1966) *The Theory of Spinors* (Dover Publications: New York)

[5] Cornwell, J.F. (1984) *Group Theory in Physics* (Academic Press: New York)

[6] Corson, E.M. (1953) *Introduction to Tensors, Spinors, and Relativistic Wave Equations* (Chelsea Publishing Company: New York)

[7] Deheuvels, R. (1993) *Tenseurs et spineurs* (Presses Universitaires de France: Paris)

[8] Dirac, P.A.M. (1928) *Proc. Roy. Soc.* A, **117**, 610–624

[9] Greiner, W. (1994) *Quantum Mechanics.* Vol. I: *An Introduction.* Vol. II: *Symmetries* (Springer-Verlag: Berlin)

[10] Hladik, J. (1994) *Le calcul tensoriel en physique* (2nd edn) (Masson: Paris)

[11] Hladik, J. (1995) *Théorie des groupes en physique et chimie quantiques* (Masson: Paris)

[12] Hladik, J. (1997) *Mécanique quantique. Atomes et molécules* (Masson: Paris).

[13] Kahan, T. (1960) *Théorie des groupes en physique classique et quantique.* Tome I: *Structures mathématiques et fondements quantiques.* Tome II: *Applications en physique classique* (Dunod: Paris)

[14] Landau L. and Lifschitz, E. (1972) *Relativistic Quantum Theory* (Pergamon: Oxford)

[15] Lévy-Leblond, J.M. (1967) *Comm. Math. Phys.*, **6**, 286

[16] Morand, M. (1973) *Géométrie spinorielle* (Masson: Paris)

[17] Penrose, R and Rindler, W. (1984) *Spinors and Space–Time* (Cambridge University Press: Cambridge)

[18] Sattinger, S.H. and Weaver, O.L. (1986) *Lie Groups and Algebras with Applications to Physics, Geometry and Mechanics* (Springer-Verlag: New York)

Index

Algebra 171
Clifford 181, 183, 179
isomorphic Lie 173
Lie 171, 172, 174, 176, 184, 189
Lie $\mathfrak{so}(2)$ 175, 176
Lie $\mathfrak{so}(3)$ 177, 184
Lie $\mathfrak{su}(2)$ 173
quaternion 181
anticommutation 21, 101, 185
anti-Hermiticity 75
anti-orthochronous 123
antisymmetric
metric tensor 29
symbol 109
associated Legendre polynomials 85

Basis
canonical 78, 80, 85
natural 25
orthonormal 28
representation 40, 210
bispinor 162

Cartan 3, 7
character tables 216
Clebsch–Gordan 150
Clifford 3
commutation 21, 75, 138, 172
commutator 129
compact 36
components of a spinor 25

Darwin 100
Dirac 3, 100, 157
direct product 130, 203
direct sum 214
of irreducible representations 215

Eigenvalues 102
of infinitesimal matrices 76
eigenvector 78
electric potential 104, 157
equation
Dirac's 157, 158, 159, 164
Pauli's 104
relativistic 161
Schrödinger's 100, 105
Euler angles 16, 44, 92
exponential mapping 45

Field
electromagnetic 104, 108, 157, 159
spinor 110

tensor 108
vector 108
four-vector 121, 147, 152, 154

Goudsmit 99
group 197
$\boldsymbol{SL}(2,\mathbb{C})$ 140, 189
Clifford 187
connected 178
continuous 35, 42
continuous linear 47
continuous point 200
definition of 200
finite 38
finite point 198
generalized Lorentz 121, 122
Lie 35, 38, 45, 48, 122, 176
linear 47
Lorentz 124, 161
orthogonal 48, 122
permutation 197
of rotations in a plane 41
of spacial rotations 36, 48, 67
product 202
proper Lorentz 135, 149
spinor 187
symmetric 198
unimodular 140
unimodular unitary $\boldsymbol{SU}(2)$ 50
unitary $\boldsymbol{U}(2)$ 47, 48
unitary unimodular 48, 187
groups
homomorphic 86, 203
isomorphic 203

Hamilton 181
relativistic 157
homomorphism 91

Infinitesimal generator 41, 42, 43, 173, 179, 183
inverse 201, 212
of a tensor product of matrices 218
isomorphic 130
spaces 22
isomorphism 22
isotropic vector 8, 22

j-spinors 54
Jacobi
identity 173

Kronecker symbol 109

Lévy-Leblond 100
Lie 43, 74, 174, 175, 176, 184, 189
bracket 172, 174

Matrix
Dirac 182
equivalent 213
exponential 73
Hermitian 181
infinitesimal 43, 45, 68, 75, 80, 128, 137, 149
infinitesimal Hermitian 45
infinitesimal rotation 18, 51
of operators 102, 207
Pauli 19, 21, 51, 148, 159, 164, 180
quaternion 188
reflection 208
regular 211, 212
rotation 14, 20, 21, 67, 137, 207, 208
transformation 12
unit 102, 109
unitary 10, 49, 218
momentum
angular orbital 110
total angular 110

Operator 205
infinitesimal 41, 82
linear 101, 203, 207
momentum vector 100
orbital momentum 111
rotation 39, 203
unit 107
unitary 107
orthochronous 123

Parameters
Cayley–Klein 14, 15
continuous 37
Olinde–Rodrigues 15
Pauli 3, 20, 99, 105, 148, 159, 164, 180

plane symmetry 9
product
 bilinear 172
 Hermitian 27, 49
 tensor of matrices 217, 218

Quaternions 181, 187

Relativistic invariance 160
relativity 121
representation 207, 210
 binary 86
 equivalent spinor 89
 irreducible 40, 87, 130, 146, 215
 matrix 40, 144, 187, 210, 212
 pure spinor 90
 reducible 40
 spinor 40, 67, 145
 standard 166
 tensor 40
 three-dimensional 52
 unitary 75
representations
 $(2j+1)$-dimensional 92
 of $\boldsymbol{SL}(2,\mathbb{C})$ 143
 equivalent 91, 175, 212, 215
 irreducible 78, 139
  of weight j 84
  of $\boldsymbol{SO}(3)$ 74, 75
  of $\boldsymbol{SU}(2)$ 55
 Lie algebra 174
 linear 39
 Lorentz group 135
 reducible 178, 214
 spinor 85
reflections 121, 122
 anti-orthochronous 123
 orthochronous 123
rotation 12, 16, 67
 'active' 67
 anti-orthochronous 123
 four-dimensional 125, 126, 135
 infinitesimal 41, 71
 Lorentz 122
 orthochronous 123
 'passive' 67
 in a plane 36, 136
 spacial 36
 in space of spinors 14
 two-dimensional 127
 of a vector field 108

Schur's lemma 55
sheet 124
space
 configuration 205
 connected 178
 Minkowski 183, 188
 representation 40, 52, 210
 vector, of spinors 24
space–time 121
spherical harmonics 82, 84, 90
spin 3, 99
 of a vector field 110
spinor
 definition 14
 dotted 141, 164
 operator 164
 of order $2j$ 54
 higher order 142
 of the Pauli equation 105
 of rank $(2j+1)(2j'+1)$ 142
 symmetric 145
 unitary 5, 7, 9, 15, 18, 28
stereographic projection 4
structure coefficients 43, 69, 173
subalgebra 182, 187
 even 180
 odd 180
subgroups 202
subspace
 complementary 214
 one-parameter 46
 orthogonal 215
 stable 40
symmetrization 144

Transformation
 general Lorentz 121
 infinitesimal 138
 Lorentz two-dimensional 136
 orthochronous Lorentz 123
 orthogonal 125
 proper Lorentz 121
 symmetry 199

Uhlenbeck 99

Vector potential 104, 157

Weight
  of irreducible representation 79, 139
  of irreducible representations 149
  of two matrices 216
  of two representations 216, 219

Zeeman 99